Palgrave Studies in Climate Resilient Societies

Series Editor
Robert C. Brears, Avonhead, New Zealand

The Palgrave Studies in Climate Resilient Societies series provides readers with an understanding of what the terms **resilience and climate resilient** societies mean; the best practices and lessons learnt from various governments, in both non-OECD and OECD countries, implementing climate resilience policies (in other words what is 'desirable' or 'undesirable' when building climate resilient societies); an understanding of what a resilient society potentially looks like; knowledge of when resilience building requires slow transitions or rapid transformations; and knowledge on how governments can create coherent, forward-looking and flexible policy innovations to build climate resilient societies that: support the conservation of ecosystems; promote the sustainable use of natural resources; encourage sustainable practices and management systems; develop resilient and inclusive communities; ensure economic growth; and protect health and livelihoods from climatic extremes.

Nyong Princely Awazi

Building Climate Resilient Communities Along Africa's Coasts

The Role of Mangroves

palgrave
macmillan

Nyong Princely Awazi ⓘ
Department of Forestry and Wildlife Technology
University of Bamenda
Bambili, Cameroon

ISSN 2523-8124   ISSN 2523-8132 (electronic)
Palgrave Studies in Climate Resilient Societies
ISBN 978-3-031-90842-2   ISBN 978-3-031-90843-9 (eBook)
https://doi.org/10.1007/978-3-031-90843-9

© The Author(s), under exclusive license to Springer Nature Switzerland AG 2025

This work is subject to copyright. All rights are solely and exclusively licensed by the Publisher, whether the whole or part of the material is concerned, specifically the rights of translation, reprinting, reuse of illustrations, recitation, broadcasting, reproduction on microfilms or in any other physical way, and transmission or information storage and retrieval, electronic adaptation, computer software, or by similar or dissimilar methodology now known or hereafter developed.
The use of general descriptive names, registered names, trademarks, service marks, etc. in this publication does not imply, even in the absence of a specific statement, that such names are exempt from the relevant protective laws and regulations and therefore free for general use.
The publisher, the authors and the editors are safe to assume that the advice and information in this book are believed to be true and accurate at the date of publication. Neither the publisher nor the authors or the editors give a warranty, expressed or implied, with respect to the material contained herein or for any errors or omissions that may have been made. The publisher remains neutral with regard to jurisdictional claims in published maps and institutional affiliations.

This Palgrave Macmillan imprint is published by the registered company Springer Nature Switzerland AG
The registered company address is: Gewerbestrasse 11, 6330 Cham, Switzerland

If disposing of this product, please recycle the paper.

Acknowledgments

The author is grateful to all those who supported this book project in one way or another. I owe a debt of gratitude to the reviewers and editors whose pertinent comments helped in fine-tuning and improving on the quality of the manuscript.

Contents

1	**Introduction**	1
	1.1 Overview of the Coastal Challenges in Africa	2
	1.2 The Role of Mangroves in Coastal Resilience in Africa	4
	1.3 Mangroves as Lifelines for Livelihoods in Africa	7
	1.4 The Interconnection of Local Communities and Mangrove Ecosystems in Africa	10
	1.5 Community-Led Initiatives and Case Studies in Africa	12
	1.6 Importance of Policy and Collaboration for Sustainable Mangrove Management in Africa	15
	1.7 The Road Ahead: Toward a Climate-Resilient Future with Mangroves in Africa	18
	1.8 Structure of the Book	20
	References	22
2	**Guardians of the Red Sea: Mangroves and Community Resilience Along the Red Sea Coast of Africa**	31
	2.1 Introduction to Africa's Red Sea Coast	32
	2.2 Ecological Importance of Mangroves Along the Rd Sea Coast of Africa	34

2.3	Dependence of Local Livelihoods on Mangroves Along Africa's Red Sea Coast	39
2.4	Case studies of Successful Mangrove Restoration Projects Along Africa's Red Sea Coast	41
2.5	Community Resilience and Adaptation Through Mangroves Along Africa's Red Sea Coast	43
2.6	Policy and Institutional Support for Mangroves Along Africa's Red Sea Coast	45
2.7	Conclusion	47
References		48

3 Central Africa's Coast: Mangrove Ecosystems and Community Resilience — 55

3.1	Introduction to Central Africa's Coastal Zones	56
3.2	Rich Biodiversity and Mangrove Ecosystem Services Along the Coasts of Central Africa	57
3.3	Traditional Knowledge and Sustainable Practices for Mangrove Ecosystems in Central Africa	61
3.4	Community-Led Initiatives in Mangrove Conservation in Central Africa	63
3.5	Industrialization, Deforestation, and the Impact on Mangroves in Central Africa	66
3.6	Resilience Building Through Mangroves and Policy Implications in Central Africa	68
3.7	Conclusion	70
References		71

4 East Africa's Coast: Bridging People and Nature for Community Resilience Through Sustainable Management of Mangrove Ecosystems — 79

4.1	Overview of East Africa's Coastal Challenges	80
4.2	Socio-Economic Impacts of Mangrove Loss in East Africa	82
4.3	Community-Based Mangrove Conservation Initiatives in East Africa	86

4.4	Bridging People and Nature: Integrating Local Knowledge with Science for Sustainable Mangrove Management in East Africa	88
4.5	Policy Recommendations for Sustainable Mangrove Management in East Africa	90
4.6	Conclusion	92
	References	93

5 Southern Africa's Coast: Mangroves for Resilience in the Face of Climate Change — 101

5.1	Introduction to Southern Africa's Coastal Regions	102
5.2	Mangrove Ecosystems for Climate Change Adaptation in Southern Africa	104
5.3	Mangroves and Local Livelihoods in Southern African Coastal Regions	110
5.4	Collaborative Efforts for Mangrove Conservation in Southern African Coastal Regions	112
5.5	Climate Change Mitigation Through Mangroves in Southern African Coastal Regions	114
5.6	Conclusion	116
	References	117

6 West Africa's Coast: Lessons in Resilience and Adaptation Through Sustainable Management of Mangrove Ecosystems — 123

6.1	Introduction to West Africa's Coastal Areas	124
6.2	The Impact of Industrialization on Mangrove Ecosystems in West Africa	126
6.3	Successful Regional Initiatives on Mangrove Conservation in West Africa and Lessons Learned	129
6.4	Sustainable Mangrove Management Practices and Community Engagement in West Africa	136
6.5	Future Directions for Mangrove Conservation in West Africa	138
6.6	Conclusion	140
	References	141

7	Conclusion	151
7.1	Key Insights	152
7.2	The Vital Role of Mangroves to Climate Resilience Along Africa's Coasts	153
7.3	Lessons from Successful Case Studies Along the Coast of Africa	154
7.4	The Socio-economic Value of Mangroves along Africa's Coasts	154
7.5	Collaborative Approaches to Sustainable Management of Mangroves Along Africa's Coasts	155
7.6	Strategic Recommendations for the Future	156
7.7	The Path Toward Building Climate-Resilient Communities Along Africa's Coasts	157
7.8	Call to Action	157
Index		159

About the Author

Dr. Nyong Princely Awazi serves as Senior Lecturer in the Department of Forestry and Wildlife Technology, College of Technology (COLTECH), the University of Bamenda, Cameroon. He holds a Ph.D. in Agroforestry and Valuation of Ecosystem Services from the University of Dschang, Cameroon. Since 2014, he has been involved in research and consultancy on the cross-cutting themes of agroforestry, forestry, ecotourism, climate change mitigation and adaptation, natural resource management and biodiversity conservation. He has consultancy experience across several countries in Africa, South America, the Caribbeans, Asia, and North America. He has (co)authored over 115 publications (72 peer reviewed articles, over 15 conference papers and posters, and over 30 book chapters) in journals and books published by reputable publishers such as Elsevier, Springer, Taylor and Francis, MDPI, Wiley, and others. He equally serves as a reviewer and till date, he has reviewed over 650 manuscripts for journals of international repute such as *Communications Earth & Environment; Sustainable Production* *and Consumption; Environmental Impact Assessment Review; Geography and Sustainability; Sustainability Science; Regional Environmental Change; Plants, People, Planet; Agroforestry Systems; Climate and Development; International Journal of Agricultural Sustainability; Sustainability; Frontiers in Forests and Global Change; Forest Science and Technology; Frontiers in Plant Science; Frontiers in Genetics; Frontiers in Sustainable Food Systems; Forest and Society.* He sits on the Editorial Boards of Frontiers in Forests and Global Change, and Frontiers in Sustainable Food Production published by Frontiers; SN Social Sciences, Discover Conservation, Discover Agriculture published by Springer Nature; and Open Journal of Forestry published by Scientific Research. He is a member of different national and international associations including the Cameroon Academy of Young Scientists (CAYS); British Ecological Society (BES) Grant Review Committee; International Union of Agroforestry (IUAF); International Union of Forest Research Organizations (IUFRO); YALI (Young African Leaders Initiative) Network; Congo Basin Young Agroforesters Association; Scholars Academic and Scientific Society (SAS) Young Research Fellow Member (SYRFM); International Society for Development and Sustainability (ISDS). He has served as Climate Change Consultant for FOKABS Canada as well as Biodiversity Conservation and Ecotourism Consultant for LEORON Institute in Riyadh, Saudi Arabia. He is equally a Post-Doctoral Research Fellow with the EU-funded TC4BE project coordinated by Wageningen University and Research—The Netherlands. He can be found on the following online research platforms:

- Academia (https://ubda.academia.edu/NyongPrincelyAwazi)
- Frontiers Loop (https://loop.frontiersin.org/people/1473281/overview)
- Google Scholar (https://scholar.google.com/citations?user=spxm5xQAAAAJ&hl=en)
- LinkedIn (https://www.linkedin.com/in/nyong-princely-awazi-552b87145/)
- ORCID (https://orcid.org/my-orcid?orcid=0000-0002-0801-0719)
- ResearchGate (https://www.researchgate.net/profile/Nyong-Princely-Awazi)

- Scopus (https://www.scopus.com/authid/detail.uri?authorId=57210996419)
- Web of Science (https://www.webofscience.com/wos/author/record/AAO-1853-2020

List of Figures

Fig. 1.1	Distribution of mangroves along the coasts of Africa (*Source* Adapted from Michel, 2014)	5
Fig. 1.2	Community-based mangrove management for improved livelihoods (*Source* Aheto et al., 2016)	9
Fig. 2.1	Mangrove distribution along the Red Sea Coast of Africa (*Source* Blanco-Sacristán et al., 2022)	35
Fig. 3.1	Map showing distribution of mangroves along the coasts of Central Africa	58
Fig. 6.1	Distribution of mangroves along the coasts of West Africa (*Source* Oyebade et al., 2010)	130

List of Tables

Table 2.1	Mangrove species along the Red Sea Coast of Africa	36
Table 3.1	Mangrove species in Central Africa	60
Table 4.1	Mangrove species in East Africa	83
Table 5.1	Mangrove species in Southern Africa	105
Table 6.1	Mangrove species diversity across countries in West Africa	131

1

Introduction

Abstract Coastal regions in Africa face increasing challenges, including erosion, habitat loss, and the impacts of climate change. Mangrove ecosystems play a crucial role in addressing these challenges by enhancing coastal resilience and providing numerous benefits to local communities. This chapter explores the significance of mangroves as lifelines for livelihoods, particularly in coastal areas where fishing, agriculture, and tourism are primary sources of income. The interconnection between local communities and mangrove ecosystems is explored, highlighting how traditional knowledge and practices contribute to the sustainable management of these vital habitats. Community-led initiatives are showcased through case studies from various African regions, demonstrating the positive outcomes of grassroots efforts in mangrove restoration and conservation. Furthermore, the chapter emphasizes the importance of policy frameworks and collaborative approaches involving local, national, and international stakeholders to ensure the long-term sustainability of mangrove ecosystems. Effective policies are critical to integrating mangrove conservation into broader coastal management strategies and addressing the challenges posed by climate change. As Africa faces an uncertain climate future, the resilience of its coastal communities depends on the continued health of mangrove ecosystems. This chapter underscores the need for a collaborative and inclusive approach to safeguarding mangroves for a climate-resilient future in Africa, ensuring that both the environment and livelihoods are protected for generations to come.

Keywords Coastal resilience · Mangrove ecosystems · Livelihoods · Community-led initiatives · Sustainable management · Climate change

1.1 Overview of the Coastal Challenges in Africa

Coastal communities in Africa face a complex array of challenges, exacerbated by both natural and human-induced pressures. With over 30 African countries possessing coastlines, these communities are particularly vulnerable to the cascading effects of climate change, including rising sea levels, intensifying storms, coastal erosion, and flooding (Akinsemolu & Olukoya, 2020; Le, 2020; Oloyede et al., 2022; Victor et al., 2025). In addition to the environmental threats, socio-economic pressures, ranging from overfishing to the impacts of tourism, agriculture, and industrial activities, further strain these regions, making them hotspots of both ecological and human vulnerability. One of the most pressing challenges faced by coastal regions in Africa is the rising sea levels caused by climate change (Nyadzi et al., 2020). Sea levels across African coasts are projected to rise by up to 1 meter by the end of the century, with significant regional variations (Ayugi et al., 2023; Nhantumbo et al., 2023). African coastlines, particularly in low-lying areas like the Nile Delta (Egypt), the Gambia River Basin, and parts of West Africa, are highly susceptible to flooding and inundation, endangering both people and ecosystems. These rising sea levels result in the salinization of freshwater resources, agricultural lands, and aquifers, further threatening food and water security (Victor et al., 2025).

Coastal erosion presents another serious challenge to Africa's coastlines. Coastal erosion, exacerbated by both climate change and unsustainable human activities, is leading to the loss of vital coastal land (Govender et al., 2025; Romaric et al., 2025). Studies have highlighted the vulnerability of countries like Ghana and Kenya, where large stretches of coastline are eroding at alarming rates, displacing communities and destroying critical infrastructure (Ideki & Ajoku, 2024; Kutor et al., 2025). This erosion also leads to the destruction of coastal habitats such as mangroves, which play a crucial role in protecting against storm surges and supporting biodiversity. Intensifying storms and flooding are also significant threats to coastal communities. Cyclones, particularly in the eastern and southern parts of the continent (like the case in Mozambique and Madagascar), have become more frequent and severe, causing widespread destruction. For example, Cyclone Idai, which struck in 2019, killed over 1,000 people and displaced millions in Mozambique, Zimbabwe, and Malawi (Mutasa, 2022). These storms not only damage infrastructure but also disrupt local economies, particularly agriculture, which is highly dependent on predictable weather patterns (Miklyaev & Olubamiro, 2025).

Beyond environmental challenges, socio-economic pressures also exacerbate the vulnerability of coastal communities in Africa. Overfishing is one of the most pressing issues, driven by the depletion of fish stocks due to unsustainable practices, illegal fishing, and the expansion of industrial fishing fleets (Mensah et al., 2025; Saidu, 2025). As a result, many coastal communities, especially those reliant on fishing as their primary livelihood, face diminished catches and economic hardship (Kakama et al., 2025; Owusu, 2025). This problem is compounded by the increasing demands of tourism, agriculture, and industrialization, all of which place additional stress on fragile coastal ecosystems. Oil exploration and other industrial activities have contributed significantly to the degradation of coastal and marine environments. In Nigeria's Niger Delta, oil spills have polluted vast stretches of coastline, leading to the destruction of fishing grounds and agricultural lands (Aa et al., 2022). These activities also impact local health and well-being, creating long-term challenges for affected communities (Okwunwa et al., 2024). Furthermore, rapid urbanization along Africa's coasts, driven by population growth and migration, has resulted in the encroachment of urban areas into coastal ecosystems, further exacerbating pressures on both the environment and local resources. The socio-economic impacts of these environmental and industrial pressures are felt most acutely by the most vulnerable populations, such as women, children, and marginalized ethnic groups. These communities often lack the resources to adapt to the changing environmental conditions, such as access to financial support or technology that could help mitigate damage from storms or rising sea levels.

Given the scale and interconnectedness of these challenges, there is a clear urgency for solutions that integrate both environmental and community resilience (Adebayo, 2024; Seçmen & Ibrahim, 2025). Strengthening the capacity of coastal communities to adapt to the challenges of climate change, coupled with sustainable development practices that reduce environmental degradation, will be essential. This includes promoting sustainable fisheries management, enhancing disaster preparedness, developing early warning systems, and advocating for the inclusion of marginalized groups in decision-making processes. Effective policy frameworks, at both national and regional levels, will also be vital in addressing these complex, interlinked issues.

1.2 The Role of Mangroves in Coastal Resilience in Africa

Mangroves are among the most productive and ecologically significant ecosystems in the world, particularly along the coastlines of Africa (Das et al., 2022; Kathiresan, 2021; Naidoo, 2023; Nwabueze, 2024). These coastal forests are found along the coasts of Africa (Fig. 1.1), where saltwater and freshwater mix provides numerous benefits to both the environment and human communities (Bunting et al., 2023). In Africa, where coastal communities are highly vulnerable to the effects of climate change, mangroves serve as vital natural buffers against a range of environmental stresses, including storms, floods, and coastal erosion. In addition to their physical protective role, mangroves contribute to biodiversity conservation, support local fisheries, and aid in carbon sequestration, further enhancing community resilience to the growing impacts of climate change.

1 Introduction 5

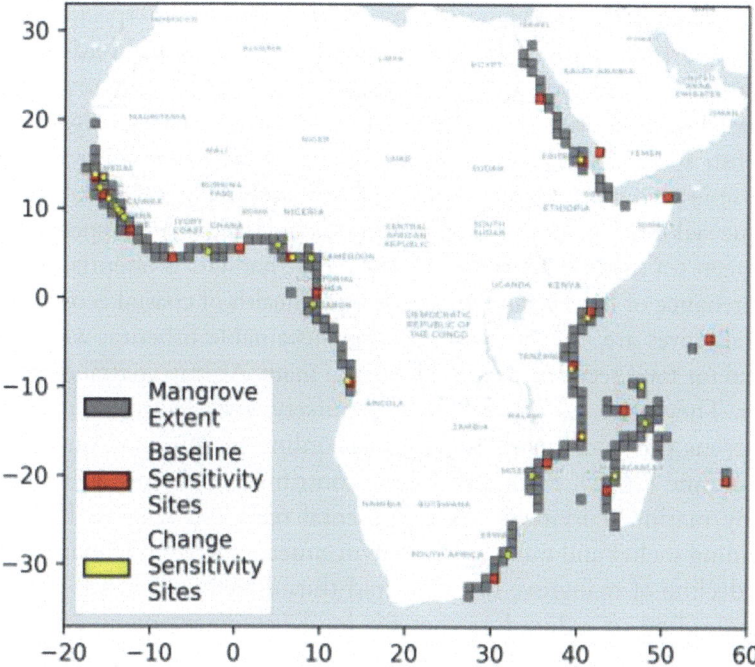

Fig. 1.1 Distribution of mangroves along the coasts of Africa (*Source* Adapted from Michel, 2014)

Ecologically, mangroves act as natural barriers that reduce the impacts of coastal storms and flooding. Their complex root systems trap sediments and stabilize shorelines, reducing the risk of erosion and the destruction of coastal infrastructure. According to Whitehead (2022) and Alves et al. (2022), mangroves are particularly effective in attenuating the force of waves, offering a protective shield for coastal settlements against the destructive power of storms and cyclones. This function is becoming increasingly important as climate change leads to rising sea levels and more intense weather events. In the context of Africa, where many coastal communities depend on fishing, agriculture, and tourism, the protection mangroves provide to infrastructure and livelihoods is crucial. The role of mangroves in maintaining biodiversity is another key aspect of their ecological value. Mangrove ecosystems are rich in species diversity, supporting a variety of fish, crustaceans, mollusks, and bird species. According to Popoola (2022) and Irabor et al. (2024), mangroves serve as vital breeding and nursery habitats for many commercially important fish species, such as shrimp and tilapia, which contribute to local economies and food security. Additionally, mangroves provide habitat for unique wildlife, including migratory birds and endangered species like the green sea turtle. The preservation of these habitats is essential for the maintenance of biodiversity and the overall health of coastal ecosystems.

Mangroves are integral to supporting sustainable fisheries, which are crucial for food security and livelihoods in many African coastal communities. They serve as critical breeding, nursery, and feeding grounds for numerous fish and shellfish species. According to Aju and Aju (2021) and Dunne (2022), mangrove forests contribute to the productivity of nearby marine ecosystems, including coral reefs and seagrass beds, by providing shelter and nutrient-rich environments for juvenile marine life. The decline of mangrove forests would thus directly affect fish populations, leading to reduced catches for local fishermen, many of whom rely on artisanal fishing for their income and sustenance. In this sense, mangroves help sustain both marine biodiversity and human communities dependent on marine resources. The link between mangroves and fisheries is particularly evident in countries such as Senegal, Mozambique, and Tanzania, where artisanal fishing is a significant economic activity. The health of mangrove ecosystems is directly correlated with the availability of marine resources, which in turn supports the livelihoods of millions of people. As these ecosystems continue to face threats from deforestation, coastal development, and pollution, it becomes essential to implement policies that promote the conservation and restoration of mangrove forests.

Another critical function of mangroves along the coasts of Africa is their ability to sequester carbon, making them valuable in the fight against climate change. Mangrove forests store carbon in both their biomass and the sediment below them, where it remains trapped for centuries. Mangroves sequester carbon at rates that are up to four times higher than tropical rainforests (Alongi, 2022), making them some of the most effective ecosystems for carbon storage. In the context of Africa, where climate change is already affecting agricultural productivity and coastal infrastructure, preserving and restoring mangrove forests can play a vital role in mitigating the impacts of greenhouse gas emissions (Aju & Aju, 2021). Through the sequestration of large amounts of carbon, mangroves contribute to global efforts to combat climate change, while also enhancing the resilience of local communities to its impacts. For instance, the restoration of mangrove ecosystems along the coasts of countries like Kenya and Ghana can help reduce the carbon footprint of these nations, while simultaneously providing a natural buffer against rising sea levels and increasing storm intensity.

Mangroves contribute significantly to the resilience of coastal communities by providing ecological, economic, and social benefits. As climate change exacerbates the risks of coastal flooding, erosion, and storm surges, the protective role of mangroves becomes more critical. By mitigating the impacts of these hazards, mangroves safeguard not only physical infrastructure but also the cultural and social fabric of coastal communities. In addition to their protective role, mangroves can generate income opportunities through ecotourism, sustainable fisheries, and carbon credit markets. According to Foli et al. (2021) and Sam et al. (2023), the restoration and sustainable management of mangrove forests along the coasts of Africa can promote economic resilience by creating job opportunities in coastal communities. Furthermore, mangroves play an essential role in enhancing food security, as they support fisheries that are a primary protein source for many African households.

1.3 Mangroves as Lifelines for Livelihoods in Africa

Mangroves, unique coastal ecosystems, play an essential role in sustaining the livelihoods of millions of people in Africa, especially those living in coastal and low-lying regions such as the case of the Volta estuary region of Ghana (Fig. 1.2). These ecosystems, which thrive in intertidal zones, provide a range of critical services that support local economies, particularly through fisheries and ecotourism. They are integral to the survival of communities that depend on these resources for food, income, and cultural practices. However, with increasing environmental degradation, the destruction of mangroves poses significant risks to these communities, highlighting the urgent need for effective restoration efforts. One of the primary ways mangroves support local livelihoods is through their contribution to fisheries. Mangrove forests serve as breeding and nursery grounds for various fish species, including commercially important ones like shrimp, tilapia, and various finfish species. According to Gnansounou et al. (2021) and Jape and Najar (2024), mangroves contribute to around 75% of the fish catch in tropical regions, including Africa, where artisanal fishing is a primary source of income and sustenance for coastal communities. The rich biodiversity within mangroves ensures a continuous supply of fish and other marine resources, which are vital not only for local food security but also for generating income through fishing. Communities that rely on mangrove-associated fisheries face significant economic losses if these ecosystems are destroyed or degraded. Moreover, mangroves support ecotourism, which has become an increasingly important economic sector in many coastal regions of Africa. The biodiversity within mangrove ecosystems, ranging from a wide variety of fish to unique bird species, makes them attractive destinations for ecotourism. In countries like Kenya, Tanzania, and Mozambique, mangrove forests have contributed to local economies by attracting tourists interested in birdwatching, boat tours, and other nature-based activities. According to Otundo Richard (2024) and Nwala and Gesiere (2024), ecotourism linked to mangroves can provide sustainable income for local communities and create employment opportunities in areas such as guiding, hospitality, and handicrafts. The integration of sustainable tourism initiatives has the potential to boost both conservation efforts and local economies simultaneously.

However, the destruction of mangroves due to human activities such as logging, coastal development, and pollution threatens these vital resources. The loss of mangroves leads to a decline in fish populations, reducing the catch available to local fishermen and affecting their income and food sources. Furthermore, the destruction of mangroves weakens coastal protection, making communities more vulnerable to the impacts of climate change, such as coastal erosion, storm surges, and rising sea levels. According to Elisha and Felix (2021), coastal erosion

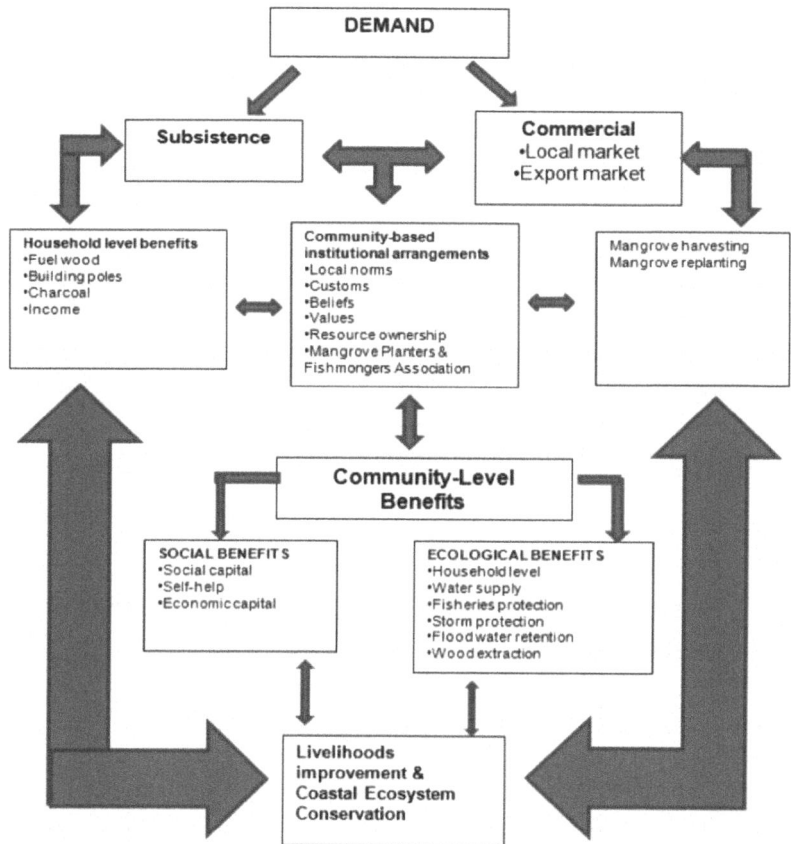

Fig. 1.2 Community-based mangrove management for improved livelihoods (*Source* Aheto et al., 2016)

resulting from mangrove loss in the Niger Delta can lead to the displacement of communities and a reduction in the availability of productive land for agriculture and other livelihoods. The loss of these ecosystems has a ripple effect on local economies, resulting in widespread poverty and food insecurity. Mangrove restoration initiatives offer significant potential for both ecological recovery and economic recovery. Restoring mangrove ecosystems can help reinstate the biodiversity and functionality of these environments, thereby supporting fisheries and ecotourism. For example, in the Gambia, efforts to restore mangrove areas have led to an increase in fish stocks and a subsequent rise in fishing income for local communities (Acosta-Alba et al., 2022; Bayo et al., 2022). Additionally, the restoration of mangroves can improve coastal resilience by providing natural barriers against storm surges and reducing the impacts of climate change, which in turn helps safeguard local livelihoods. Mangrove restoration projects also create employment opportunities for local populations, contributing to the development of sustainable livelihoods. For example, community-based restoration programs that involve planting mangrove saplings or protecting existing mangrove stands can offer jobs in restoration activities, environmental monitoring, and ecotourism development. By engaging local communities in restoration efforts, these initiatives can foster a sense of ownership and responsibility for the conservation of mangrove ecosystems.

1.4 The Interconnection of Local Communities and Mangrove Ecosystems in Africa

Mangrove ecosystems, found along the coasts of Africa, play a critical role in the ecological health of the continent (Akram et al., 2023; Santos et al., 2024). These ecosystems provide numerous benefits, such as protecting coastal areas from erosion, serving as nurseries for fish species, and sequestering carbon (Gnansounou et al., 2022). However, their existence and preservation are also deeply intertwined with the lives of local communities who rely on these environments for sustenance, livelihood, and cultural practices. The relationship between African local communities and mangrove ecosystems is symbiotic, with both the ecosystem and the communities depending on each other for survival and prosperity.

For centuries, local communities across Africa have coexisted with mangrove ecosystems, using their resources in a sustainable manner. According to studies by Agbekpornu et al. (2021), Chuku et al. (2022), and Senghor et al. (2023), many communities in West Africa, including those in Ghana and Senegal, have employed traditional knowledge and practices to manage mangroves. These include selective harvesting, where only certain trees are cut down, and the preservation of mangrove areas during their reproductive seasons to ensure regeneration. In East Africa, coastal communities in Tanzania and Kenya have similarly relied on local knowledge to harvest wood, shellfish, and other resources without depleting the mangrove resources (Nyangoko et al., 2022). The management practices, passed down through generations, are deeply rooted in customs and beliefs that foster a sense of responsibility and stewardship toward the environment. Local leaders and elders often play a key role in maintaining these traditions, ensuring that the younger generation understands the importance of conserving mangroves for both their survival and the community's long-term well-being. These practices have created a sustainable system that, for the most part, has allowed mangrove ecosystems to thrive alongside human populations.

Local ecological knowledge (LEK) is a crucial tool in the conservation and management of mangrove ecosystems in Africa. As discussed by Albuquerque et al. (2021) and Brondízio et al. (2021), LEK refers to the understanding that local communities have developed over time about their environment, which includes an awareness of the interrelationships between species, ecosystem processes, and seasonal patterns. In mangrove ecosystems, such knowledge enables communities to predict environmental changes, such as tidal fluctuations, rainfall patterns, and the growth cycles of mangrove species, and adapt their practices accordingly. For instance, in the coastal regions of Guinea-Bissau, traditional fishermen use their knowledge of mangrove tidal systems to determine the optimal fishing times, ensuring that they harvest only when fish stocks are abundant and that they do not overexploit vulnerable fish species (Davies-Vollum et al., 2024; Ramos, 2022). This practice not only sustains local livelihoods but also maintains the integrity of mangrove ecosystems by preventing overfishing. However, in recent decades, external pressures such as urbanization, industrial development, and climate change have threatened the health of mangrove ecosystems in Africa, necessitating the integration of modern conservation techniques alongside traditional practices. This integration ensures that mangrove management remains effective in the face of evolving environmental challenges.

To achieve long-term success in the conservation of mangrove ecosystems, there has been an increasing effort to combine traditional knowledge with modern conservation methods in Africa. A notable example is the work done by the African Mangrove Network (AMN) and the Cameroon Mangroves and Wetlands Network (CMN) in partnership with local communities in countries like Mozambique and Cameroon (CMN, 2024; FAO, 2009) AMN has facilitated workshops where local community members collaborate with conservationists to share knowledge and adapt modern scientific approaches to their traditional practices. For example, remote sensing and GIS (Geographical Information System) technology are used to monitor mangrove health and deforestation rates, which helps communities track the impact of their resource use and take proactive measures when necessary (Mohammed et al., 2025; Numbere, 2022). At the same time, local knowledge about sustainable harvesting practices and seasonal cycles is integrated into these monitoring efforts, ensuring that the conservation strategies are culturally appropriate and practically applicable. An essential component of this integration is the empowerment of local communities to take ownership of mangrove conservation. As noted by Adeyanju et al. (2021) and Yanou et al. (2023), community-based natural resource management (CBNRM) initiatives that combine indigenous knowledge and scientific research tend to yield more sustainable outcomes. In the context of mangrove ecosystems, when local communities are given the tools and knowledge to manage their natural resources, they are more likely to adopt conservation practices that benefit both the environment and their livelihoods.

1.5 Community-Led Initiatives and Case Studies in Africa

Africa's coastal regions are home to some of the world's most significant mangrove ecosystems. Over the past few decades, community-led mangrove restoration and conservation initiatives have emerged as effective models for preserving these ecosystems (Ravaoarinorotsihoarana et al., 2023). These initiatives are rooted in the understanding that local populations, who depend on mangroves for fishing, timber, and other resources, have the most at stake and are thus best positioned to ensure the long-term sustainability of these ecosystems.

In West Africa, particularly in Senegal, the *Centre de Suivi Ecologique* (CSE) has been pivotal in implementing community-driven mangrove restoration projects along the Saloum River Delta (CSE, 2023). The initiative, known as the "Mangrove Restoration Project," has empowered local communities, particularly women, to engage in the rehabilitation of degraded mangrove forests (Medina et al., 2023). Through training and the provision of resources, these communities have not only restored more than 1,000 hectares of mangrove forest but have also gained increased livelihoods through the sustainable harvesting of mangrove resources such as honey, crabs, and fish. Mangrove restoration projects in Africa have demonstrated that local stewardship, combined with scientific knowledge, leads to enhanced ecological and socio-economic outcomes (Keleman et al., 2024). The success of this initiative highlights the importance of a participatory approach, where local stakeholders are involved from the planning stages through to implementation and monitoring.

In East Africa, Kenya's coastal communities have similarly led efforts to restore mangrove forests. One notable example is the initiative led by the *Kenya Marine and Fisheries Research Institute* (KMFRI) in partnership with local community groups (Kiprono, 2021). This initiative focuses on the replanting of mangrove species such as *Rhizophora mucronata* and *Avicennia marina* in the regions of Lamu and Kwale. Local communities have been actively engaged in the restoration process, planting seedlings and taking part in long-term monitoring activities. The community-led approach has had remarkable results, increasing mangrove cover and improving local fish populations, which are essential for the communities' food security and economic well-being (Tschentscher et al., 2023). This success has been attributed to the involvement of community members in both the management and decision-making processes, fostering a sense of ownership and responsibility over the restoration efforts.

Further south, in Mozambique, the provinces of Cabo Delgado, Inhambe, Maputo, Nampula, Sofala, and Zambezia have witnessed exemplary cases of community-led mangrove conservation. The *Mozambique Mangrove Conservation and Restoration Initiative* in these provinces, led by local NGOs in collaboration with the government, focus on the rehabilitation of overexploited mangrove forests (High Commission of the Republic of Mozambique, 2022). Local communities in these provinces are involved in both the restoration of degraded areas and the sustainable use of mangrove resources. A study by Macamo et al. (2024) found that community engagement in monitoring and enforcing sustainable mangrove harvesting practices resulted in the regeneration of large areas of mangrove forest, as well as improved community resilience to climate change impacts. The lessons of mangroves conservation and restoration from these provinces in Mozambique underscore the importance of community involvement in resource management as a pathway to environmental and socio-economic benefits.

In Central Africa, particularly in Gabon, Cameroon, and Equatorial Guinea, local communities have been working alongside environmental organizations such as the World Wildlife Fund (WWF) to restore mangrove habitats (WWF, 2018). The Cameroon Mangroves and Wetlands Network (CMN) focuses on enhancing the ecological health of mangroves while addressing the needs of local communities that rely on these ecosystems for fuelwood and other resources. Through this network, communities are encouraged to adopt sustainable harvesting practices and participate in mangrove monitoring programs, ensuring the long-term health of the ecosystem. The integration of traditional ecological knowledge with modern scientific approaches has proven crucial for the success of mangrove conservation in Cameroon (Ajonina, 2022; Bissonnette et al., 2024; Grimm et al., 2024).

The case studies from Senegal, Kenya, Mozambique, Gabon, Cameroon, and Equatorial Guinea illustrate the broad potential of community-led mangrove restoration efforts across Africa. The common thread in all these initiatives is the recognition that local communities, when empowered, can serve as effective stewards of their environment. These efforts have led to not only ecological improvements, such as increased mangrove cover and biodiversity, but also enhanced community resilience to climate change and the creation of alternative livelihoods that reduce pressure on the mangrove ecosystem. These initiatives also highlight the importance of incorporating local knowledge and involving communities in decision-making processes, as it fosters a deeper sense of ownership and responsibility over conservation efforts. The success of these community-led initiatives suggests that similar models can be replicated across the African continent to restore degraded mangrove ecosystems. Lessons learned from these case studies include the need for strong local participation, the integration of scientific research with local knowledge, and the provision of sustainable livelihoods. With continued support for such community-driven efforts, Africa's mangrove ecosystems can be restored, protected, and managed for future generations.

1.6 Importance of Policy and Collaboration for Sustainable Mangrove Management in Africa

Mangroves are vital ecosystems that provide numerous benefits, including coastal protection, biodiversity conservation, and carbon sequestration. In Africa, where coastal communities are heavily reliant on the resources provided by mangroves, sustainable management of these ecosystems is crucial. However, mangrove forests are facing significant threats from human activities, such as deforestation, land reclamation, and pollution, along with climate change impacts. The importance of policy and collaboration in ensuring sustainable mangrove management cannot be overstated, as it forms the foundation for long-term conservation and restoration efforts.

A comprehensive policy framework is essential to ensure the protection and sustainable use of mangroves. In many African countries, the lack of clear policies and regulations for mangrove management has contributed to their degradation (Mohamed et al., 2023). Therefore, national governments must develop and implement policies that promote sustainable mangrove conservation while balancing the needs of local communities who depend on these resources for their livelihoods. These policies should focus not only on mangrove protection but also on promoting sustainable alternatives for resource use. A key aspect of effective mangrove policy is the integration of mangrove management into broader national development strategies. As noted by Ferreira et al. (2022), development policies that fail to account for environmental concerns often lead to unsustainable practices that harm ecosystems like mangroves. In this context, policies must be designed to align economic, social, and environmental goals. This could include incentivizing sustainable fisheries, ecotourism, and sustainable agriculture, which can reduce pressure on mangrove forests and enhance local economies. For example, the use of market-based approaches, such as payments for ecosystem services (PES), could be explored to create financial incentives for mangrove conservation.

Effective governance is a critical determinant of successful mangrove management. Governance involves the establishment of clear roles and responsibilities for various stakeholders, the enforcement of policies, and the creation of mechanisms for accountability and transparency. In many African countries, weak governance structures, lack of institutional capacity, and limited enforcement of laws have contributed to mangrove degradation (Mwanja et al., 2024). Strengthening institutional capacity at both the national and local levels is vital for ensuring that policies are implemented effectively. Moreover, governance should be inclusive, involving all relevant stakeholders including governments, non-governmental organizations (NGOs), local communities, and the private sector. Empowering local communities to manage and protect mangroves can be a key strategy, as local knowledge and traditional practices often align with sustainable conservation (Brownson et al., 2024). The implementation of community-based management models can provide both ecological and socio-economic benefits, ensuring that the people who are most affected by mangrove degradation are actively involved in decision-making processes.

Mangroves are often located in transboundary coastal areas, making regional cooperation and international collaboration essential for their protection and management. African countries must work together to develop and implement joint strategies that transcend national borders. International cooperation provides an opportunity to share knowledge, resources, and best practices, which can be particularly valuable in regions where mangrove ecosystems span several countries, such as the Gulf of Guinea and the East African coast (Hapres, 2022). The most significant frameworks for regional cooperation are the Nairobi and Abidjan Conventions, which provides a platform for African coastal countries to collaborate on the sustainable management of marine and coastal resources, including mangroves. Through these Conventions, African countries have adopted regional action plans that outline strategies for the conservation and restoration of mangroves, as well as protocols for sharing information and research (Hamukuaya, 2024). Additionally, international agreements such as the Ramsar Convention on Wetlands, which includes mangrove ecosystems, can support Africa's efforts to conserve these critical habitats.

A key challenge in mangrove conservation is the tension between environmental protection and economic development. As highlighted by Naidoo (2023), the conversion of mangrove forests for agriculture, aquaculture, and urban expansion is often driven by the need for economic growth. However, the long-term benefits of mangrove ecosystems, such as their role in reducing coastal erosion and providing fish nursery habitats, far outweigh the short-term economic gains from their destruction. Therefore, integrating environmental considerations into national and regional development strategies is essential for ensuring the sustainable use of mangroves. Governments must adopt a more integrated approach to development that recognizes the value of ecosystem services provided by mangroves. This can be achieved by incorporating environmental impact assessments into decision-making processes, promoting green infrastructure, and ensuring that mangrove conservation is a priority in development agendas.

1.7 The Road Ahead: Toward a Climate-Resilient Future with Mangroves in Africa

Africa's coastal regions are among the most vulnerable to climate change, facing rising sea levels, stronger storms, and increased flooding. These pressures threaten ecosystems, livelihoods, and economies, especially for communities that depend on coastal resources for survival. As the continent grapples with these challenges, it is becoming increasingly evident that nature-based solutions, such as mangrove restoration, can play a pivotal role in building climate resilience. Mangroves, with their unique ability to store carbon, protect coastal communities from storm surges, and sustain biodiversity, offer a promising pathway to mitigate and adapt to climate change. However, the full potential of these ecosystems can only be realized through the scaling up of community-based restoration initiatives and their integration into broader climate adaptation and mitigation strategies. Empowering local communities with the necessary knowledge and tools to manage their ecosystems sustainably is central to fostering long-term, self-sustaining resilience.

Mangroves are an extraordinary group of coastal trees that thrive in the brackish waters of tropical and subtropical regions. They provide a range of ecosystem services that are crucial for climate resilience. Their dense root systems stabilize shorelines, preventing erosion and reducing the impact of storm surges and tidal waves, which have become increasingly destructive due to climate change (Kochoni et al., 2023). Additionally, mangroves are highly effective carbon sinks, sequestering carbon at rates higher than most terrestrial forests. This carbon storage capacity makes mangroves an essential component of global climate mitigation efforts (Kiribou et al., 2024). Moreover, mangrove ecosystems support a rich diversity of species, contributing to both biodiversity conservation and local fisheries, which many coastal African communities rely on for food and income.

Despite their significance, mangrove ecosystems across Africa have been significantly degraded over the past few decades. According to a study by Naidoo (2023), Africa lost over 50% of its mangrove cover between 1980 and 2020, primarily due to unsustainable development practices, urbanization, and deforestation for fuelwood and timber. The loss of mangroves has exacerbated vulnerability to climate-related events, such as flooding, coastal erosion, and reduced fish stocks. This trend of mangrove loss not only diminishes the ecosystems' ability to provide climate resilience but also undermines the livelihoods of millions of people who depend on them. Given this context, the urgency of scaling up successful community-based mangrove restoration initiatives cannot be overstated. Local communities, as the primary stewards of these ecosystems, play a crucial role in the restoration and sustainable management of mangroves. Community-led restoration projects have proven effective in many African countries, demonstrating how local knowledge and participation can lead to successful outcomes. For example, the restoration of mangroves in Senegal's Casamance region has been led by local fishermen, who have re-established mangrove forests, improving coastal protection and fish stocks (Cormier-Salem, 2024; Mbaye et al., 2022). These initiatives are vital because they provide not only ecological benefits but also economic opportunities for local communities. Empowering these communities with the right tools, knowledge, and resources is essential for scaling up restoration efforts. Climate resilience can be built through education and capacity-building programs that enable local communities to understand the importance of mangroves and the best practices for their restoration. It is also essential to provide access to financial and technical support for mangrove projects, ensuring that communities can maintain these efforts over the long term. In many cases, the integration of mangrove restoration with broader climate adaptation and mitigation strategies such as community-based early warning systems, sustainable fisheries management, and climate-smart agriculture can maximize the benefits of these initiatives. Moreover, engaging local governments, NGOs, and the private sector is crucial to creating a supportive enabling environment for these efforts.

To move toward a climate-resilient future, Africa must not only restore mangrove ecosystems but also ensure that these efforts are integrated into national and regional climate policies. The African Union's Agenda 2063 and the Paris Agreement provide frameworks for advancing climate adaptation and mitigation on the continent, and mangrove restoration must be a key component of these strategies. Strengthening regional collaborations and sharing knowledge on best practices for mangrove restoration will further enhance the effectiveness of these efforts.

1.8 Structure of the Book

This book has seven (07) chapters, with chapter 1 being the introduction laying emphasis on coastal challenges in Africa, the role of mangroves in coastal resilience in Africa, mangroves as lifelines for livelihoods in Africa, the interconnection of local communities and mangrove ecosystems in Africa, community-led initiatives and case studies in Africa, importance of policy and collaboration for sustainable mangrove management in Africa, and mangroves for a climate-resilient future in Africa. Chapter 2 focuses on mangroves and community resilience along the Red Sea Coast of Africa presenting an overview of the Red Sea's geographical and socio-economic importance, along with climate change challenges and mangroves' role in the region; describes the biodiversity supported by mangroves and their role in coastal stability, storm protection, and supporting local fisheries; discusses the relationship between local livelihoods (fishing, agriculture, ecotourism) and the health of mangrove ecosystems; highlights community-driven restoration efforts and successful partnerships in Egypt, Sudan, and Eritrea; explores traditional knowledge and local adaptation strategies using mangroves to address climate challenges; reviews policy frameworks and regional cooperation, offering recommendations for strengthening mangrove conservation efforts; and summarizes lessons learned from the Red Sea region, stressing the need for integrated, community-focused solutions. Chapter 3 focuses on mangrove ecosystems and community resilience along central Africa's coast providing an overview of coastal zones in Cameroon and Gabon, focusing on ecological and socio-economic importance, and challenges like industrialization; discusses the biodiversity of mangrove forests and their vital role in supporting fisheries and carbon sequestration; explores indigenous practices and the integration of traditional knowledge with modern conservation efforts; highlights successful restoration projects and the empowerment of local communities in Cameroon and Gabon; discusses the impact of industrialization and deforestation on mangrove ecosystems and strategies for mitigating these effects; reviews resilience-building strategies and the role of policy in sustainable mangrove management; and summarizes key lessons from Central Africa, emphasizing the need for integrated, multi-stakeholder approaches. Chapter 4 lays emphasis on bridging people and nature for community resilience through sustainable management of mangrove ecosystems along East Africa's coast providing an overview of Kenya, Tanzania, and Somalia's coastal challenges, including overfishing and land degradation, and the role of mangroves; examining the effects of mangrove deforestation on local communities, economies, and climate vulnerability; presenting examples of successful community-based mangrove restoration efforts across East Africa; discusses the importance of blending traditional knowledge with scientific research to enhance mangrove conservation; provides policy recommendations for national and regional cooperation to enhance mangrove protection; and a conclusion which emphasizes the importance of bridging people and nature for effective mangrove conservation and community resilience. Chapter 5 focuses on the role of mangroves in the face of climate change along Southern Africa's coast, presenting an overview of coastal ecosystems in Mozambique, South Africa, and Namibia, focusing on climate change threats and mangrove protection; explains how mangroves mitigate climate risks and enhance resilience to storms and flooding; discusses the economic value of mangroves and the impact of climate change on livelihoods; highlights cross-border cooperation and successful mangrove restoration efforts; examines the role of mangroves in climate change mitigation through carbon storage; and summarizes Southern Africa's approach to mangrove conservation, emphasizing integrated strategies for climate adaptation. Chapter 6 lays emphasis on lessons in resilience and adaptation through sustainable management of mangrove ecosystems along West Africa's coast presenting an overview of coastal regions in Senegal, Nigeria, and Ghana, discussing key challenges like oil exploration and urbanization; discusses the effects of industrialization on mangrove habitats and the socio-economic consequences of mangrove loss; case studies of successful mangrove restoration and regional collaborations for mangrove protection; examines the role of local communities in managing mangrove ecosystems and strategies for sustainable use; policy recommendations for improving mangrove protection and scaling up successful initiatives; and highlights key lessons from West Africa and the importance of integrating mangrove conservation into sustainable development. Chapter 7 is the conclusion which recaps key findings from the previous chapters, reinforcing the vital role of mangroves in building climate resilience; expatiates on the multifaceted role of mangroves in enhancing environmental and community resilience; highlights key takeaways from successful mangrove restoration projects across Africa; discusses the economic importance of mangroves and their role in sustainable livelihoods; emphasizes the importance of cross-sectoral partnerships for effective mangrove conservation; provides actionable strategies for scaling up mangrove conservation efforts; stresses the interconnectedness of people and nature in building resilient communities through mangrove protection; and a call to action for all stakeholders to prioritize mangrove conservation as part of climate resilience efforts in Africa.

References

Aa, I., Op, A., Ujj, I., & Mt, B. (2022). A critical review of oil spills in the Niger Delta aquatic environment: Causes, impacts, and bioremediation assessment. *Environmental Monitoring and Assessment, 194*(11), 816.

Acosta-Alba, I., Nicolay, G., Mbaye, A., Dème, M., Andres, L., Oswald, M., Avadí, A., et al. (2022). Mapping fisheries value chains to facilitate their sustainability assessment: Case studies in The Gambia and Mali. *Marine Policy, 135*, Article 104854.

Adebayo, W. G. (2024). Resilience in the face of ecological challenges: Strategies for integrating environmental considerations into social policy planning in Africa. *Sustainable Development*. https://doi.org/10.1002/sd.3113

Adeyanju, S., O'connor, A., Addoah, T., Bayala, E., Djoudi, H., Moombe, K., Sunderland, T., et al. (2021). Learning from community-based natural resource management (CBNRM) in Ghana and Zambia: Lessons for integrated landscape approaches. *International Forestry Review, 23*(3), 273–297.

Agbekpornu, H., Ennin, E. J., Issah, F., Pappoe, A., & Yeboah, R. (2021). Women in West African Mangrove Oyster (Crassostrea Tulipa) harvesting, contribution to food security and nutrition in Ghana. *Oceanography and Fisheries, 14*(1), 001–0019.

Aheto, D. W., Kankam, S., Okyere, I., Mensah, E., Osman, A., Jonah, F. E., & Mensah, J. C. (2016). Community-based mangrove forest management: Implications for local livelihoods and coastal resource conservation along the Volta estuary catchment area of Ghana. *Ocean & Coastal Management, 127*, 43–54.

Ajonina, G. N. (2022). Cameroon Mangroves: Current status, uses, challenges, and management perspectives. In *Mangroves: biodiversity, livelihoods and conservation* (pp. 565–609). Springer Nature Singapore.

Aju, P. C., & Aju, J. A. (2021). Mangrove forests in Nigeria: Why their restoration, rehabilitation and conservation matters. *African Journal of Environmental and Natural Science Research, 4*(1), 84–93.

Akinsemolu, A. A., & Olukoya, O. A. (2020). The vulnerability of women to climate change in coastal regions of Nigeria: A case of the Ilaje community in Ondo State. *Journal of Cleaner Production, 246*, Article 119015.

Akram, H., Hussain, S., Mazumdar, P., Chua, K. O., Butt, T. E., & Harikrishna, J. A. (2023). Mangrove health: A review of functions, threats, and challenges associated with mangrove management practices. *Forests, 14*(9), 1698.

Albuquerque, U. P., Ludwig, D., Feitosa, I. S., De Moura, J. M. B., Gonçalves, P. H. S., Da Silva, R. H., Ferreira Júnior, W. S., et al. (2021). Integrating traditional ecological knowledge into academic research at local and global scales. *Regional Environmental Change, 21*, 1–11.

Alongi, D. M. (2022). Impacts of climate change on blue carbon stocks and fluxes in mangrove forests. *Forests, 13*(2), 149.

Alves, R. B., Bapentire, A. D., Almar, R., Louarn, A., Rossi, P. L., Corsini, L., & Morand, P. (2022). *Compendium: Coastal management practices in West Africa: Existing and potential solutions to control coastal erosion, prevent flooding and mitigate damage to society*. The World Bank. https://www.documentation.ird.fr/hor/fdi:010085569

Ayugi, B. O., Chung, E. S., Zhu, H., Ogega, O. M., Babousmail, H., & Ongoma, V. (2023). Projected changes in extreme climate events over Africa under 1.5° C, 2.0° C and 3.0° C global warming levels based on CMIP6 projections. *Atmospheric Research, 292*, 106872.

Bayo, B., Habib, W., & Mahmood, S. (2022). Spatio-temporal assessment of mangrove cover in the Gambia using combined mangrove recognition index. *Advanced Remote Sensing, 2*(2), 74–84.

Bissonnette, J. F., Dossa, K. F., Nsangou, C. A., Satchie, Y. A., Moussa, H., Miassi, Y. E., Onguene, R., et al. (2024). What occurs within the mangrove ecosystems of the douala region in cameroon? Exploring the challenging governance of readily available woody resources in the Wouri Estuary. *Environments, 11*(6), 121.

Brondízio, E. S., Aumeeruddy-Thomas, Y., Bates, P., Carino, J., Fernández-Llamazares, Á., Ferrari, M. F., Shrestha, U. B., et al. (2021). Locally based, regionally manifested, and globally relevant: Indigenous and local knowledge, values, and practices for nature. *Annual Review of Environment and Resources, 46*(1), 481–509.

Brownson, S. U., Chigbu, G., & Osazuwa, C. M. (2024). Cultural security and environmental conservation: Exploring the link between indigenous knowledge systems and sustainable resource management in cross rivers state. *The American Journal of Management and Economics Innovations, 6*(08), 13–40.

Bunting, P., Hilarides, L., Rosenqvist, A., Lucas, R. M., Kuto, E., Gueye, Y., & Ndiaye, L. (2023). Global mangrove watch: Monthly alerts of mangrove loss for Africa. *Remote Sensing, 15*(8), 2050.

Chuku, E. O., Effah, E., Adotey, J., Abrokwah, S., Adade, R., Okyere, I., Crawford, B., et al. (2022). Spotlighting women-led fisheries livelihoods toward sustainable coastal governance: The estuarine and mangrove ecosystem shellfisheries of West Africa. *Frontiers in Marine Science, 9*, Article 884715.

CMN. (2024). *Cameroon mangroves & Wetlands network: Towards a more significant position amongst actors in wetlands ecosystem conservation and more effective actions for sustainable development in municipalities.* https://www.cmrmangroves-wetlands.net/

Cormier-Salem, M. C. (2024). Desirable futures: Perspectives of Joola fisherwomen in Casamance, Senegal. *Futures, 162*, Article 103435.

CSE. (2023). *Projet « Suivi des Risques côtiers et solutions douces au Bénin, Sénégal et Togo».* https://www.cse.sn/?page_id=14054

Das, S. C., Thammineni, P., & Ashton, E. C. (2022). Mangroves: A unique ecosystem and its significance. In S.C. Das, Pullaiah, & E. C. Ashton (Eds.), Mangroves: Biodiversity, livelihoods and conservation. Singapore. https://doi.org/10.1007/978-981-19-0519-3_1

Davies-Vollum, K. S., Koomson, D., & Raha, D. (2024). Coastal lagoons of West Africa: A scoping study of environmental status and management challenges. *Anthropocene Coasts, 7*(1), 7.

Dunne, A. (2022). Nutrition and organism flows through tropical marine ecosystems. https://doi.org/10.25781/KAUST-0BJ13

Elisha, O. D., & Felix, M. J. (2021). Destruction of coastal ecosystems and the vicious cycle of poverty in Niger Delta region. *Journal of Global Agriculture and Ecology, 11*(2), 7–24.

FAO. (2009). *The African Mangrove Network (AMN)*. https://www.fao.org/4/ak995e/ak995e06.pdf

Ferreira, A. C., Borges, R., & de Lacerda, L. D. (2022). Can sustainable development save mangroves? *Sustainability, 14*(3), 1263.

Foli, B. A. K., Williams, I. K., Boakye, A. A., Azumah, D. M. Y., Agyekum, K. A., & Wiafe, G. (2021). Earth observation services in support of West Africa's blue economy: Coastal resilience and climate impacts. *Remote Sensing in Earth Systems Sciences*, 1–12.

Gnansounou, S. C., Salako, K. V., Sagoe, A. A., Mattah, P. A. D., Aheto, D. W., & Glèlè Kakaï, R. (2022). Mangrove ecosystem services, associated threats and implications for wellbeing in the Mono Transboundary Biosphere Reserve (Togo-Benin), West-Africa. *Sustainability, 14*(4), 2438.

Gnansounou, S. C., Toyi, M., Salako, K. V., Ahossou, D. O., Akpona, T. J. D., Gbedomon, R. C., Kakaï, R. G., et al. (2021). Local uses of mangroves and perceived impacts of their degradation in Grand-Popo municipality, a hotspot of mangroves in Benin, West Africa. *Trees, Forests and People, 4*, Article 100080.

Govender, I. H., Reddy, M., & Pillay, R. P. (2025). *A review of residual flood risks in South African-vulnerable coastal communities: Opportunities to influence policy*. https://doi.org/10.5772/intechopen.1008977. https://www.intechopen.com/online-first/1207389

Grimm, K., Spalding, M., Leal, M., Kincaid, K., Aigrette, L., Amoah-Quiminee, P., Zimmer, M., et al. (2024). *Including local ecological knowledge (LEK) in Mangrove conservation & restoration*. A Best-Practice Guide for Practitioners and Researchers.

Hamukuaya, H. (2024). *Assessment of transboundary environmental issues affecting biodiversity in shared marine ecosystems: Towards formulating harmonised regional framework for conservation of aquatic biodiversity and joint action plan*. http://repository.au-ibar.org/bitstream/handle/123456789/1594/Final%20Afr%20Mar%20BiO%20Assessments%203RD%20%20March%2c%202023%20DR.%20HASHALI%20REPORT.pdf?sequence=1&isAllowed=y

Hapres, L. (2022). Africa blue economy strategies integrated in planning to achieve sustainable development at national and regional economic communities (RECs). *Journal of Sustainability Research, 4*(3), e220011. https://doi.org/10.20900/jsr20220011

High Commission of the Republic of Mozambique. (2022). *Mozambique to restore 185,000 hectares of mangrove forests.* https://www.mozambiquehighcommission.org.uk/mozambique-to-restore-185-000-hectares-of-mangrove-forests.html

Ideki, O., & Ajoku, O. (2024). Scenario analysis of shorelines, coastal erosion, and land use/land cover changes and their implication for climate Migration in East and West Africa. *Journal of Marine Science and Engineering, 12*(7), 1081.

Irabor, A. E., Obakanurhe, O., Ozor, A. O., Adagha, O., Sanubi, J. O., Chukwurah, A. I., Zelibe, S. A., et al. (2024). Is small-scale fishing sustainable in Delta State, Nigeria? A glance into the problems and possible solutions. *Fisheries Research, 274*, Article 106981.

Jape, K. K., & Najar, M. A. (2024). Empowering local stewardship of coastal ecosystems in Zanzibar: Participatory models for habitat rehabilitation and resilience. *International Journal of Advanced Multidisciplinary Research, 4*(2), 185–194.

Kakama, G. A., Pétursson, J. G., Kristófersson, D. M., Ibengwe, L. J., & Tómasson, T. (2025). Role of microfinance to improve livelihoods of small Scale Inland and coastal fisheries communities in Africa. In *Handbook of Sustainable Blue Economy* (pp. 1–26). Springer Nature Switzerland.

Kathiresan, K. (2021). Mangroves: Types and importance. In R. P. Rastogi, M. Phulwaria, & D. K. Gupta (Eds.), *Mangroves: Ecology, biodiversity and management*. Springer. https://doi.org/10.1007/978-981-16-2494-0_1

Keleman, P. J., Sá, R. M., & Temudo, M. P. (2024). Drifting away from the roots: Genderfluidity as Diola's mangrove fishing strategies in three island-villages of Northern Guinea-Bissau. *Human Ecology, 52*(5), 935–951.

Kiprono, A. (2021). *An assessment of the effectiveness of mangrove restoration projects along the Kenyan coast* (Doctoral dissertation, University of Nairobi). https://erepository.uonbi.ac.ke/bitstream/handle/11295/155712/Kiprono%20_An%20Assessment%20of%20the%20Effectiveness%20of%20Mangrove%20Restoration%20Projects%20Along%20the%20Kenyan%20Coast.pdf?sequence=1&isAllowed=y

Kiribou, R., Djene, S., Bedadi, B., Ntirenganya, E., Ndemere, J., & Dimobe, K. (2024). Urban climate resilience in Africa: A review of nature-based solution in African cities' adaptation plans. *Discover Sustainability, 5*(1), 94.

Kochoni, B. I., Avakoudjo, H. G. G., Kamelan, T. M., Sinsin, C. B. L., & Kouamelan, E. P. (2023). Contribution of mangroves ecosystems to coastal communities' resilience towards climate change: A case study in southern Cote d'Ivoire. *GeoJournal, 88*(4), 3935–3951.

Kutor, S. K., Ofori, O. D., Akyea, T., & Arku, G. (2025). Climate change-immobility nexus: Perspectives of voluntary immobile populations from three coastal communities in Ghana. *Climatic Change, 178*(2), 1–23.

Le, T. D. N. (2020). Climate change adaptation in coastal cities of developing countries: Characterizing types of vulnerability and adaptation options. *Mitigation and Adaptation Strategies for Global Change, 25*(5), 739–761.

Macamo, C. D. C. F., Inácio da Costa, F., Bandeira, S., Adams, J. B., & Balidy, H. J. (2024). Mangrove community-based management in Eastern Africa: Experiences from rural Mozambique. *Frontiers in Marine Science, 11*, 1337678.

Mbaye, A. A., Lefèvre, F. G., Sarr, A., Sambou, C., Gueye, A., Gueye, F., Sarr, K. Y., Araba, A. C., Gaye, C. A. B., & Dieng, M. (2022). *A situational analysis of small-scale fisheries in Senegal: From vulnerability to viability—Challenges and opportunities for fisheries governance* (V2V Working Paper 2022-2). V2V Global Partnership, University of Waterloo, Canada.

Medina, L., Krendelsberger, A., Renkamp, T., Madurga-Lopez, I., Pacillo, G., Läderach, P., Hellin, J., Sieber, S., & Bonatti, M. (2023). *Community voices on climate, peace and security: Senegal*. CGIAR FOCUS Climate Security. 66 p. https://hdl.handle.net/10568/137319

Mensah, J., Amoah, J. O., Amenumey, E. B., Mattah, P. A., & Mattah, M. M. (2025). The closed fishing season policy in Ghana: Lessons from the coastal fisherfolk. *Marine Policy, 171*, Article 106472.

Michel, J. (2014). Oil spills in mangroves; planning & response considerations. Technical Report, 96p. https://response.restoration.noaa.gov/sites/default/files/Oil_Spill_Mangrove.pdf

Miklyaev, M., & Olubamiro, O. C. (2025). *Impacts of climate change and hazards on key sectors in Madagascar* (No. 2025-01). JDI Executive Programs. https://cri-world.com/publications/qed_dp_4626.pdf

Mohamed, M. K., Adam, E., & Jackson, C. M. (2023). Policy review and regulatory challenges and strategies for the sustainable mangrove management in Zanzibar. *Sustainability, 15*(2), 1557.

Mohammed, A. H., Salem, M., Farg, E., & Mohamed, S. A. (2025). Earth observation data for mangrove monitoring and management at the red sea coastline, Egypt. In *Modelling and advanced earth observation technologies for coastal zone management* (pp. 145–175). Springer Nature Switzerland.

Mutasa, C. (2022). Revisiting the impacts of tropical cyclone Idai in Southern Africa. In *Climate impacts on extreme weather* (pp. 175–189). Elsevier.

Mwanja, R. I., Peter, K. H., & Said, M. K. (2024). The efficacy of management practices in combating mangrove forest degradation: A case study from Pangani District-Tanzania. *East African Journal of Science, Technology and Innovation, 6*.

Naidoo, G. (2023). The mangroves of Africa: A review. *Marine Pollution Bulletin, 190*, Article 114859.

Nhantumbo, B. J., Dada, O. A., & Ghomsi, F. E. (2023). *Sea level rise and climate change-impacts on African coastal systems and cities*. https://doi.org/10.5772/intechopen.113083. https://www.intechopen.com/online-first/88259

Numbere, A. O. (2022). Application of GIS and remote sensing towards forest resource management in mangrove forest of Niger Delta. In *Natural resources conservation and advances for sustainability* (pp. 433–459). Elsevier.

Nwabueze, B. C. (2024). *A research framework to assess the contribution of the mangrove ecosystem to the well-being of coastal communities in Africa* (Doctoral dissertation, University of British Columbia). https://open.library.ubc.ca/media/stream/pdf/24/1.0447769/4

Nwala, P., & Gesiere, L. (2024). Tourism development in niger delta and its impact on soft power diplomacy of Nigeria 1960–2024. *Cascades, Journal of the Department of French & International Studies, 2*(2), 103–117.

Nyadzi, E., Bessah, E., & Kranjac-Berisavljevic, G. (2020). Taking stock of climate change induced sea level rise across the West African Coast. *Environmental Claims Journal, 33*(1), 77–90.

Nyangoko, B. P., Berg, H., Mangora, M. M., Shalli, M. S., & Gullström, M. (2022). Local perceptions of changes in mangrove ecosystem services and their implications for livelihoods and management in the Rufiji Delta, Tanzania. *Ocean & Coastal Management, 219*, Article 106065.

Okwunwa, C., Okeudo, G., Dike, D., Ikeogu, C., & Uzoho, M. (2024). Investigation of the socio economic impact of oil spill remediation on the health of people in oil bearing localities in the Niger Delta. *Emerald International Journal of Scientific and Contemporary Studies, 6*(1), 18–43.

Oloyede, M. O., Williams, A. B., Ode, G. O., & Benson, N. U. (2022). Coastal vulnerability assessment: A case study of the Nigerian coastline. *Sustainability, 14*(4), 2097.

Otundo Richard, M. (2024). *Nexus between strategic sustainable tourism investment projects and sustainable growth and development of Kenya's coast region*. Nexus Between Strategic Sustainable Tourism Investment Projects and Sustainable Growth and Development of Kenya's Coast Region

(September 22, 2024). https://papers.ssrn.com/sol3/papers.cfm?abstract_id=4964247

Owusu, V. (2025). Effect of rising fuel prices on small-scale fisheries livelihoods and marine sustainability in Ghana. *PLoS ONE, 20*(1), Article e0317260.

Popoola, O. M. (2022). Fish production and biodiversity conservation: An interplay for life sustenance. In: Chibueze Izah, S. (Eds.), *Biodiversity in Africa: Potentials, threats and conservation. Sustainable Development and Biodiversity* (Vol. 29). Singapore. https://doi.org/10.1007/978-981-19-3326-4_11

Ramos, C. S. (2022). *Factors influencing bivalve collection by local populations in the Bijagós Archipelago, Guinea-bissau* (Master's thesis, Universidade de Aveiro (Portugal)). https://www.proquest.com/openview/670a8c46deaaec3d1a0ddfe78e2268a8/1?pq-origsite=gscholar&cbl=2026366&diss=y

Ravaoarinorotsihoarana, L. A., Maltby, J., Glass, L., Oates, J., Rakotomahazo, C., Randrianandrasaziky, D. A., Lavitra, T., et al. (2023). Lessons for ensuring continued community participation in a mangrove blue carbon conservation and restoration project in Madagascar. *Western Indian Ocean Journal of Marine Science, 22*(2), 43–60.

Romaric, G. A., Jean-Marc, Z. B. G., & Vital, G. T. A. (2025). Impact of climatic and anthropogenic factors on the spatio-temporal dynamics of mangroves in the Grand-Bassam Wetland, Côte d'Ivoire. *International Journal of Environment and Climate Change, 15*(2), 74–86.

Saidu, M. (2025). Contributions of fisheries and aquaculture to food security in Africa. In *Food Security, Nutrition and sustainability through aquaculture technologies* (pp. 493–502). Cham: Springer Nature Switzerland.

Sam, K., Zabbey, N., Gbaa, N. D., Ezurike, J. C., & Okoro, C. M. (2023). Towards a framework for mangrove restoration and conservation in Nigeria. *Regional Studies in Marine Science, 66*, Article 103154.

Santos, M. M., Ferreira, A. V., & Lanzinha, J. C. G. (2024). *Climate change, coastal vulnerability, and mangrove protection in Africa.* Proceedings of the international conference on changing cities VI: Spatial, design, landscape, heritage & socio-economic dimensions Rhodes Island, Greece; June 24–28, 2024. https://www.researchgate.net/profile/Michael-Santos-17/publication/381854480_Climate_Change_Coastal_Vulnerability_and_Mangrove_Protection_in_Africa/links/66ed7999fc6cc4648965a8a7/Climate-Change-Coastal-Vulnerability-and-Mangrove-Protection-in-Africa.pdf

Seçmen, S., & Ibrahim, F. A. (2025). Rethinking sustainable development goals (SDG) for floating slums in African coastal settings: Makoko community in Nigeria. *Cities, 159*, Article 105751.

Senghor, K., Partelow, S., Herrera, C. G., & Osemwegie, I. (2023). Conflicting governance realities: Aligning historical and cultural practices with formal marine protected area co-management in Senegal. *Marine Policy, 155*, Article 105706.

Tschentscher, T., Chege, N., & Remaury, H. (2023). An integrated seascape approach to revitalise ecosystems and livelihoods in shimoni-vanga, Kenya. *Maiko Nishi*, 179.

Victor, N. O., Ileberi, E., Daniel, L. O., & Sinneh, I. S. (2025). Climate change impact on coastal groundwater salinity along the african coast: With model prediction of economic cost. In *Artificial intelligence and data science for sustainability: Applications and methods* (pp. 347–378). IGI Global Scientific Publishing.

Whitehead, A. J. (2022). *Green Coastal protection and flood defence options for western Indian Ocean countries* (Doctoral dissertation, Stellenbosch: Stellenbosch University). https://scholar.sun.ac.za/bitstreams/61be54b8-67ea-4fa1-9472-42d94b31aeb9/download

WWF (2018). *WWF assesses changes in mangroves forests.* https://cameroon.panda.org/?25162/WWF-assesses-changes-in-mangroves-forests

Yanou, M. P., Ros-Tonen, M. A., Reed, J., Moombe, K., & Sunderland, T. (2023). Integrating local and scientific knowledge: The need for decolonising knowledge for conservation and natural resource management. *Heliyon, 9*(11).

2

Guardians of the Red Sea: Mangroves and Community Resilience Along the Red Sea Coast of Africa

Abstract The Red Sea is a vital maritime route and a region of immense ecological and socio-economic significance. However, it faces escalating environmental challenges, including coastal erosion, overfishing, and the impacts of climate change, which threaten its biodiversity and the livelihoods of coastal communities. Mangrove ecosystems along the Red Sea serve as essential natural barriers, providing critical protection against storms, stabilizing coastal sediments, and supporting marine biodiversity. Despite their ecological importance, these mangroves are under threat from urbanization, industrialization, and unsustainable practices. This study explores the role of mangroves in enhancing community resilience along the Red Sea coast, focusing on their ecological benefits, including coastal protection and fishery support. The paper also examines local livelihoods dependent on mangrove resources, such as fishing, agriculture, and ecotourism, highlighting the socio-economic value of mangroves. Successful case studies of community-driven mangrove restoration projects in Egypt, Sudan, and Eritrea are analyzed to uncover best practices and lessons for enhancing resilience. Furthermore, the study delves into the role of traditional knowledge and community-led adaptation strategies in managing mangrove ecosystems, emphasizing the need for stronger policy support and regional cooperation. In conclusion, this study underscores the importance of integrating local communities into mangrove conservation efforts and strengthening policy frameworks to ensure the long-term resilience of coastal ecosystems and the livelihoods that depend on them.

Keywords Red Sea · Mangroves · Coastal resilience · Community adaptation · Ecosystem services · Climate change

2.1 Introduction to Africa's Red Sea Coast

The Red Sea, an essential water body that separates northeastern African from the Arabian Peninsula, is a region of considerable geographical and socio-economic significance. Stretching approximately 2,300 kilometers, the Red Sea links the Mediterranean Sea to the Gulf of Aden and the Arabian Sea, making it a crucial maritime route for international trade and commerce. Historically, the Red Sea has been central to the cultural and economic exchanges between Africa, the Middle East, and Asia, with ports such as Alexandria, Jeddah, and Massawa playing pivotal roles in regional connectivity. Its coastal regions are home to diverse ecosystems, rich marine life, and a series of small-scale economies relying heavily on fisheries and tourism. However, the Red Sea faces growing environmental and socio-economic challenges that threaten both its biodiversity and the livelihoods of millions of people who depend on its resources (Paruğ et al., 2024).

One of the most pressing challenges facing the Red Sea coast is coastal erosion. The combination of natural factors, including sea-level rise due to climate change and the increasing intensity of storms, exacerbates the loss of coastline and destruction of critical habitats. Coastal erosion has been particularly severe in countries like Egypt and Sudan, where tourist destinations and vital infrastructure are threatened. Additionally, overexploitation of marine resources (especially overfishing) has led to the depletion of fish stocks, disrupting local economies and food security (Moussalli & Feidi, 2010). The Red Sea's marine ecosystems, including coral reefs and fish populations, are under considerable stress, partly due to industrial fishing practices, pollution, and the tourism industry's environmental footprint. Climate change, further intensifying these challenges, poses another significant threat to the region. As global temperatures rise, the Red Sea is experiencing increased sea surface temperatures, which harm sensitive marine species, including corals. These changes not only affect the biodiversity of the region but also reduce the resilience of ecosystems, leaving them vulnerable to diseases and bleaching events (Genevier et al., 2019; Osman et al., 2018). The altering of rainfall patterns and the exacerbation of desertification are also direct consequences of climate change that affect the agricultural productivity and water resources of coastal communities (Elasha, 2010). Moreover, the ongoing rise in sea levels is projected to inundate low-lying coastal areas, causing displacement and increasing poverty rates in vulnerable communities.

In addressing these environmental challenges, mangroves play a critical role in the Red Sea's coastal ecosystems. Mangroves, known for their ability to thrive in saline coastal waters, act as natural buffers against coastal erosion by stabilizing sediment and reducing the impact of waves. Mangroves equally provide critical habitats for fish, crustaceans, and other marine life, promoting biodiversity and supporting local fisheries (Elmogy, 2024; Tesfamichael, 2012). The root systems of mangrove trees also absorb carbon, contributing to climate change mitigation efforts. Despite their significance, mangrove forests along the Red Sea face threats from urban development, industrialization, and the degradation of coastal ecosystems. Conservation efforts are essential to safeguard these vital habitats and the socio-economic benefits they provide to local populations. Thus, the Red Sea's coast is a dynamic region with substantial economic, environmental, and social value. However, it faces significant challenges due to coastal erosion, resource overexploitation, and the effects of climate change. The protection and restoration of mangrove forests can play an essential role in addressing these challenges, preserving the delicate balance of the region's ecosystems, and supporting the sustainable development of coastal communities.

2.2 Ecological Importance of Mangroves Along the Rd Sea Coast of Africa

Mangrove ecosystems are widely distributed along the Red Sea coast of Africa (Fig. 2.1), and they play a crucial role in maintaining ecological balance, providing biodiversity hotspots, and protecting coastal communities. These unique ecosystems (made up of different species of mangroves), characterized by salt-tolerant trees and shrubs, support a diverse array of flora and fauna, contributing significantly to regional biodiversity (Table 2.1). According to Pilcher and Alsuhaibany (2020), Red Sea mangroves harbor over 130 species of fish, 140 species of invertebrates, and several bird species, including the endangered flamingos. This biodiversity is essential for the region's ecological health, as mangroves serve as nurseries for many marine species, providing shelter and food for juvenile fish and invertebrates, which later contribute to the local and global food web. In addition to their role in supporting biodiversity, mangroves act as natural barriers against storms and flooding. Their complex root systems help to dissipate wave energy, significantly reducing the impact of coastal erosion and storm surges. This function is especially important in the context of climate change, as rising sea levels and increased storm intensity threaten coastal regions. Mangroves reduce the velocity of incoming waves and act as buffers, protecting coastal infrastructure and human populations from the devastating effects of extreme weather events. As highlighted by Khalil (2015), Alhassan and Aljahdali (2021), and Moustafa et al. (2023), the presence of mangrove forests decreases the impact of floods by stabilizing sediments and preventing soil loss.

Mangroves also play a critical role in maintaining coastal stability. The root systems of mangrove trees bind sediments together, stabilizing the coastline and preventing erosion. This is especially important along the Red Sea coast, where high salinity and fluctuating water levels make coastal ecosystems more vulnerable. The roots also facilitate the filtration of sediments and pollutants, maintaining water quality and supporting the overall health of the marine environment. According to Abdel-Hamid et al. (2018) and Haseeba et al. (2025), mangrove forests along the Red Sea coast are integral in maintaining coastal ecosystem services,

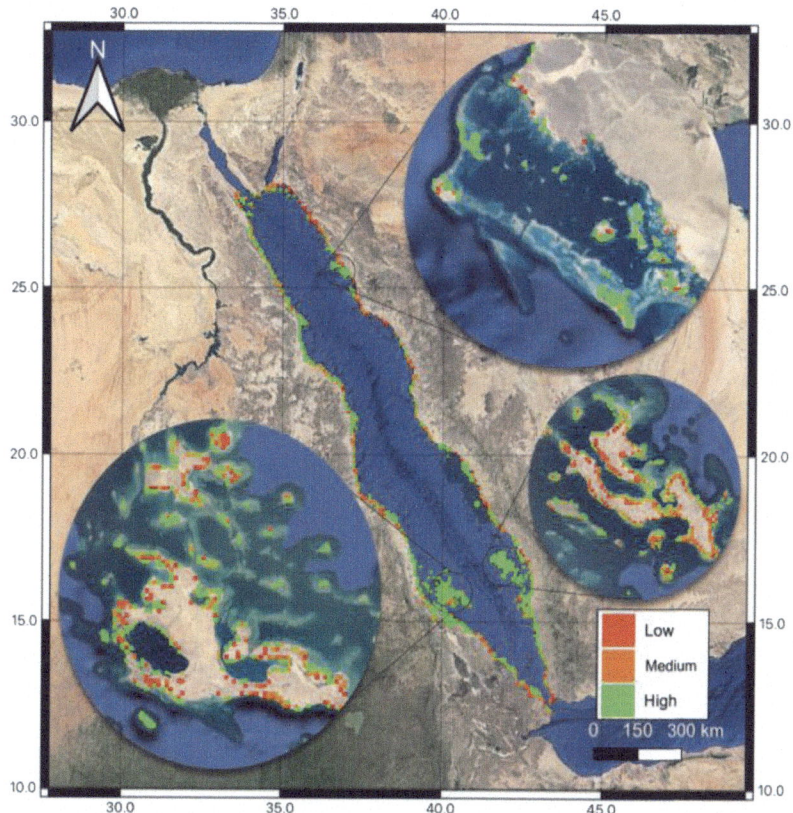

Fig. 2.1 Mangrove distribution along the Red Sea Coast of Africa (*Source* Blanco-Sacristán et al., 2022)

such as water filtration, nutrient cycling, and sediment retention. Last but not the least, mangroves contribute to fish populations by providing a unique habitat for numerous marine species. As nurseries for juvenile fish, mangrove ecosystems support the growth of species like groupers, snappers, and several species of shrimp, which are vital for local fisheries. The mangrove root structures provide shelter from predators, while the surrounding waters offer abundant food resources. As a result, healthy mangrove ecosystems directly support sustainable fish populations and the livelihoods of communities reliant on fishing.

Table 2.1 Mangrove species along the Red Sea Coast of Africa

Mangrove species	Description	References
Avicennia marina (White mangrove)	*Avicennia marina*, also known as the white mangrove, is a highly prevalent and dominant mangrove species in the Red Sea area. It is distinguished by its tolerance to high salinity and its unique grayish-green leaves, which are covered with salt-excreting glands. This species is also recognized for its pneumatophores (specialized aerial roots that extend above the soil), to enable gas exchange in low-oxygen environments. In the Red Sea, *Avicennia marina* plays a vital role in protecting the shoreline and minimizing coastal erosion. Its extensive root system helps trap sediments while offering a habitat for various marine species. Additionally, this mangrove is known for its ability to withstand environmental stresses, contributing significantly to the ecological health of mangrove forests in the region	Dicks (1986), Diop et al. (2002), and Afefe et al. (2019)

Mangrove species	Description	References
Rhizophora mucronata (Red mangrove)	*Rhizophora mucronata*, commonly known as the red mangrove, is a key species along the Red Sea coastline. It is easily recognizable by its unique stilt roots, which rise above the sediment and help anchor the tree in the constantly changing intertidal environment. These roots are essential for providing support to the mangrove and promoting sediment buildup. The red mangrove plays a vital role in protecting coastal areas, as its stilt roots form an intricate network that absorbs wave energy and reduces the effects of storm surges and erosion. Furthermore, the red mangrove creates a vital habitat for various marine species, including fish and crustaceans, which seek shelter within its dense root system	Shaltout et al. (2005), Ahmed and Abdel-Hamid (2007), and Khalil (2015)

(continued)

Table 2.1 (continued)

Mangrove species	Description	References
Ceriops tagal (Spurred mangrove)	*Ceriops tagal*, also called the spurred mangrove, is less abundant than *Avicennia marina* and *Rhizophora mucronata* but still plays an important role in the mangrove ecosystems along the Red Sea coast. This species is characterized by its knee-like aerial roots and small, rounded leaves. *Ceriops tagal* is highly adapted to the saline and nutrient-deficient conditions of the Red Sea mangrove forests. It enhances the ecological diversity of the mangrove by offering an additional habitat for both marine and terrestrial species. Its root system aids in sediment stabilization and shoreline erosion prevention, supporting the overall health of the mangrove ecosystem	Manohar et al. (2023) and Naidoo (2023)
Bruguiera gymnorrhiza (Large-leaved mangrove)	*Bruguiera gymnorrhiza*, commonly known as the large-leaved mangrove, is a species found in the Red Sea mangroves, though it is less common than other mangrove varieties. This species is easily identified by its large, tough leaves and unique, spreading aerial roots. The large-leaved mangrove is notable for its tolerance to fluctuating salinity levels and its preference for more sheltered environments compared to other mangrove species. Ecologically, *Bruguiera gymnorrhiza* plays a vital role by offering habitat for numerous marine species and assisting in the stabilization of coastal sediments. Its extensive root system helps reduce the effects of tidal movements and storms on the mangrove ecosystem	Allen and Duke (2006)

The mangrove species along the Red Sea Coast of Africa are diverse, each playing a vital role in maintaining ecological balance and coastal protection (Table 2.1). *Avicennia marina* (White Mangrove) is the most widespread species, known for its tolerance to high salinity and grayish-green leaves with salt-excreting glands. Its pneumatophores facilitate gas exchange, and its dense root system helps stabilize the shoreline and trap sediments, aiding in erosion control (Afefe et al., 2019; Dicks, 1986; Diop et al., 2002). *Rhizophora mucronata* (Red Mangrove) features distinctive stilt roots that provide stability in intertidal zones and support sediment accumulation. Its root system absorbs wave energy, mitigating coastal erosion and supporting diverse marine life (Ahmed & Abdel-Hamid, 2007; Khalil, 2015; Shaltout et al., 2005). *Ceriops tagal* (Spurred Mangrove) has knee-like aerial roots and small leaves, thriving in saline and nutrient-poor conditions. This species adds to ecological diversity, stabilizes sediments, and supports shoreline integrity (Manohar et al., 2023; Naidoo, 2023). Lastly, *Bruguiera gymnorrhiza* (Large-Leaved Mangrove) is characterized by large, leathery leaves and spreading roots. It is less common but crucial for providing habitat and stabilizing coastal sediments (Allen & Duke, 2006). These mangrove species together contribute significantly to the resilience of the Red Sea coastal ecosystems.

2.3 Dependence of Local Livelihoods on Mangroves Along Africa's Red Sea Coast

Mangroves along Africa's Red Sea coast play a crucial role in supporting local livelihoods through activities such as fishing, agriculture, and tourism. These ecosystems provide a variety of services essential for the economic well-being of coastal communities. The intricate relationship between these human activities and mangrove ecosystems underscores the significance of these habitats for both environmental and economic stability. Fishing is one of the primary livelihoods dependent on mangroves. These ecosystems act as nursery grounds for many marine species, including fish, crabs, and mollusks, which are vital for the local fishing industry. According to Waleed et al. (2024), mangroves are considered essential for supporting biodiversity, providing shelter for juvenile fish and enhancing fish populations. Coastal communities along Africa's Red Sea rely heavily on these marine resources, with mangrove forests contributing to food security and income generation. Furthermore, the ecological health of mangroves directly influences the abundance and diversity of marine species, which in turn affects the success of local fisheries. In addition to fishing, mangroves also support agriculture in many coastal areas. The nutrient-rich sediments from mangrove wetlands provide fertile soil for crops such as rice, vegetables, and fruits. Agriculture is often interlinked with small-scale aquaculture, where mangrove areas are used for pond farming. According to Little (2018), the mangrove ecosystem's ability to buffer against coastal erosion also indirectly protects agricultural lands from saltwater intrusion. These lands provide not only food for local consumption but also income for local farmers who trade surplus crops in local markets. Ecotourism is another growing economic activity linked to mangrove ecosystems. The beauty of mangrove forests, their biodiversity, and their role in coastal protection create significant potential for nature-based tourism. Mangrove-based ecotourism, such as bird watching, boat tours, and nature walks, has been identified as a promising sector for sustainable economic development in coastal regions. Studies such as those by Shaalan (2005), Madkour (2015), and Lakhouit (2019) suggest that the tourism potential of mangrove areas is largely untapped along the Red Sea coast, with many sites offering opportunities for sustainable ecotourism. Not only does this bring economic benefits to local communities, but it also raises awareness about the importance of mangroves for both environmental and human health.

However, mangrove ecosystems face significant threats from degradation, largely due to human activities such as overfishing, coastal development, and unsustainable agricultural practices. This degradation has a direct impact on the livelihoods of local communities. As mangroves are destroyed, the coastal protection they provide against storms and erosion diminishes, leaving agricultural lands and infrastructure vulnerable to natural disasters. The decline in fish populations due to mangrove loss further threatens food security and income for local fishermen. According to Madkour and Mohammed (2008), Afefe et al. (2021), and Cziesielski et al. (2021), mangrove loss has been rapid in many areas of the Red Sea due to both natural and anthropogenic pressures. The risks to local livelihoods caused by mangrove degradation are compounded by the challenges of maintaining sustainable tourism practices. As mangrove ecosystems degrade, their attractiveness as tourist destinations diminishes, reducing the potential for ecotourism to contribute to local economies. Additionally, the challenges of balancing conservation efforts with development objectives remain significant, with communities often caught between the economic benefits of resource extraction and the long-term sustainability of mangrove ecosystems.

2.4 Case studies of Successful Mangrove Restoration Projects Along Africa's Red Sea Coast

Mangrove ecosystems along Africa's Red Sea coast have long been vital for coastal protection, biodiversity, and the livelihoods of local communities. However, overexploitation, coastal development, and climate change have led to the degradation of these important habitats. In response, several successful mangrove restoration projects in Egypt, Sudan, and Eritrea have demonstrated the power of community-driven efforts, as well as the crucial roles of partnerships between local communities, non-governmental organizations (NGOs), and governments.

In Egypt, the mangrove restoration effort along the Red Sea coast has involved significant community participation. A notable project is the Red Sea Coastal Rehabilitation Program (RSCRP), which began in 2008. Local fishermen and community members were actively engaged in the restoration of mangroves in the Wadi El Gemal National Park, an area that had seen extensive mangrove loss due to urban expansion and unsustainable agricultural practices (Sarhan & Elmahdy, 2023; Sector, 2006). The project's success hinged on the involvement of local communities in both the restoration and long-term monitoring phases.

By involving local people in these activities, the program created a sense of ownership and responsibility for the mangrove ecosystem. Partnerships were crucial in this initiative. The Egyptian Environmental Affairs Agency (EEAA) partnered with NGOs, such as the Nature Conservation Egypt (NCE), to facilitate the project. These partnerships helped provide technical expertise, funding, and advocacy for mangrove protection. Through such collaborations, the RSCRP successfully restored over 400 hectares of mangroves, and the local community saw improvements in both biodiversity and fish stocks, which are vital to their livelihoods (Cziesielski et al., 2021).

Sudan has also seen promising results from community-driven mangrove restoration projects. One example is the Community-Based Environmental Management Program (CBEMP) along the Sudanese Red Sea coast. CBEMP aims to restore mangrove forests and increase local resilience to climate change by enhancing the ecological and socio-economic benefits of these ecosystems. Through partnerships between local communities, NGOs like the Sudanese Environment Conservation Society (SECS), and the government, mangrove restoration has been integrated into community development planning (USAID, 2012). The CBEMP employed a collaborative approach, with local fishermen, women, and youth participating in activities such as planting mangroves, monitoring ecological health, and developing sustainable livelihood options. These efforts have not only led to the restoration of several hectares of mangrove forests but also contributed to a reduction in coastal erosion and an increase in fish populations, which benefits the local fishing industry (Sugga et al., 2024).

In Eritrea, the Zoba Gash Barka Mangrove Restoration Project, implemented with the support of the United Nations Development Programme (UNDP), focused on restoring mangroves along the Gash River Delta, a region severely impacted by deforestation and saltwater intrusion (CBD, 2019). This project was unique in that it involved a combination of restoration and community-based conservation practices aimed at improving the livelihoods of rural coastal populations. The Gash Barka project adopted a participatory approach, working closely with local communities, including farmers and fishers, who rely on the mangroves for firewood, fish, and materials for constructing boats.

Collaboration between Eritrean environmental authorities, the UNDP, and local communities helped restore nearly 500 hectares of mangrove forests. Moreover, the project established sustainable harvesting practices and ecotourism initiatives that offered alternative income sources, reducing the pressure on mangrove resources (Awimbo et al., 2004; Gebreyohanns, 2006).

From these case studies, several key lessons and best practices emerge. Firstly, community involvement is essential for the long-term success of mangrove restoration efforts. When local people are involved in decision-making and restoration activities, they are more likely to protect and manage the ecosystem sustainably. Secondly, multi-stakeholder partnerships between governments, NGOs, and local communities are crucial for providing the necessary resources, expertise, and advocacy to ensure the success of such projects. Equally, restoring the ecological functions of mangroves often goes hand in hand with improving local livelihoods. Integrating mangrove restoration with sustainable livelihoods, such as ecotourism or sustainable fishing practices, ensures that communities see tangible benefits from their conservation efforts. Finally, adaptive management and continuous monitoring are critical in assessing the health of restored mangrove ecosystems and adjusting strategies as needed.

2.5 Community Resilience and Adaptation Through Mangroves Along Africa's Red Sea Coast

Mangroves along Africa's Red Sea coast provide critical ecological, economic, and cultural services. These coastal ecosystems, thriving in the intertidal zones of salty waters, are vital not only for biodiversity conservation but also for enhancing community resilience and adaptation to climate change. The intricate relationship between these ecosystems and local communities is grounded in both traditional knowledge and contemporary adaptation strategies aimed at safeguarding the environment and livelihoods.

Indigenous and local knowledge plays a crucial role in the sustainable management of mangrove ecosystems. In many Red Sea coastal communities, this knowledge has been passed down through generations, guiding practices that promote mangrove conservation. Traditional knowledge often incorporates an understanding of tidal patterns, plant growth cycles, and the relationship between mangroves and marine species, which are fundamental for the survival of fishing communities (Afefe, 2021; Diop et al., 2002). For instance, local fishermen often observe the seasonal variations in mangrove health to inform their fishing activities, ensuring the balance between ecosystem service extraction and conservation. In addition to this, traditional rituals and customs around mangrove harvesting are common across the region. In Eritrea, communities observe specific taboos and seasonal restrictions to allow mangrove regeneration (Kitula, 2022). These practices align with modern environmental conservation techniques, emphasizing the importance of integrating traditional ecological knowledge with scientific research to manage coastal zones effectively.

Mangroves along Africa's Red Sea coast are increasingly recognized as a crucial tool in adapting to climate change. Rising sea levels, extreme weather events, and coastal erosion are among the major threats faced by these communities, and mangroves provide an effective natural buffer. The dense root systems of mangroves help to stabilize coastlines, reduce erosion, and mitigate the impacts of storm surges, which are expected to intensify with climate change (Blanco-Sacristán et al., 2022). Moreover, mangroves sequester carbon at rates higher than most terrestrial forests, contributing to climate change mitigation (Awad et al., 2023; Youssef et al., 2024). This makes them not only a solution for adaptation but also a tool for reducing the region's carbon footprint. The restoration of degraded mangrove areas, coupled with active protection, offers a sustainable way for coastal communities to enhance their resilience. Such restoration efforts, coupled with governmental and international support, can mitigate environmental degradation and bolster community livelihoods.

For mangroves to continue providing these critical services, building community capacity is essential. Education, awareness, and participatory management are fundamental in ensuring long-term sustainable use of mangrove ecosystems. Programs that empower local communities, such as those led by the UN Environment Programme (UNEP) and local NGOs, have demonstrated success in training coastal populations in sustainable resource use and ecosystem-based adaptation (Afefe, 2021). Capacity-building initiatives, such as the creation of local mangrove management committees, enable communities to monitor the health of mangrove ecosystems and make informed decisions regarding harvesting and restoration activities. These programs also promote alternative livelihoods, such as ecotourism and sustainable aquaculture, which reduce pressure on mangrove resources while diversifying income sources (Mehanzel, 2012). Additionally, integrating women and youth into these processes ensures that knowledge sharing and leadership are inclusive, further strengthening community resilience (Hariri et al., 2000). When communities are engaged in the management of their natural resources, they are more likely to invest in the long-term health of the environment and develop strategies to cope with climate-related challenges.

2.6 Policy and Institutional Support for Mangroves Along Africa's Red Sea Coast

Mangroves along Africa's Red Sea coast are critical ecosystems that contribute to shoreline protection, biodiversity conservation, and carbon sequestration. Despite their importance, mangroves in this region face threats from coastal development, overharvesting, pollution, and climate change. Addressing these challenges requires robust policy and institutional frameworks that support mangrove conservation and sustainable management.

At the national level, several countries along Africa's Red Sea coast have recognized the importance of mangroves in their environmental policies. In Sudan, the government has implemented policies aimed at protecting coastal ecosystems, including mangroves. These policies emphasize the need for sustainable use and conservation of natural resources, with specific attention to mangrove forests (Beyer et al., 2015; Nasr et al.,

2015; Siddig et al., 2018). Similarly, in Eritrea, the government has developed strategies for sustainable coastal zone management, which include the protection and rehabilitation of mangrove areas (Haile, 1999; Ogbazghi, 2018). In Egypt, the government has adopted various measures to protect the Red Sea's coastal environments. Mangroves are listed as a protected species in Egypt's Protected Areas Law, and efforts have been made to integrate mangrove conservation into broader coastal management programs. However, there is limited enforcement of these regulations, which often hampers their effectiveness (Diop et al., 2002; Gladstone et al., 1999). Despite these efforts, the region lacks a comprehensive and coordinated approach to mangrove conservation. Policies tend to be fragmented, with insufficient resources allocated for implementation and monitoring. Furthermore, local communities, who are key stakeholders in mangrove conservation, are often excluded from decision-making processes.

Regional cooperation plays a crucial role in addressing the transboundary nature of coastal and marine resource management. The Red Sea and Gulf of Aden are shared by several countries, and collaborative efforts are essential for the sustainable management of mangrove ecosystems. The Regional Organization for the Conservation of the Environment of the Red Sea and Gulf of Aden (PERSGA) is a key player in fostering regional cooperation. PERSGA has been involved in initiatives that promote the conservation of mangroves, such as the establishment of marine-protected areas and regional monitoring programs (PERSGA, 2024). International frameworks also provide significant support for mangrove conservation in the region. The Ramsar Convention on Wetlands, which Egypt, Sudan, and Eritrea are signatories to, emphasizes the importance of wetland ecosystems, including mangroves. Additionally, the United Nations Convention on Biological Diversity (CBD) encourages countries to implement policies that protect biodiversity, including mangrove forests. Through such international agreements, countries in the region are guided toward sustainable management practices for mangroves.

To enhance mangrove conservation along Africa's Red Sea coast, several recommendations can be made. Firstly, national policies should be harmonized to create a unified approach to mangrove management. This includes the development of comprehensive coastal zone management plans that prioritize mangrove protection and integrate local community participation. Efforts should also be made to increase funding and resources for enforcement and monitoring of existing policies. Secondly, regional cooperation should be strengthened. Building on the successes of PERSGA, countries along the Red Sea should expand collaborative efforts to share knowledge, data, and best practices for mangrove conservation. This could include joint research projects and coordinated restoration efforts across borders. Thirdly, greater emphasis should be placed on the role of local communities in mangrove conservation. Policies should empower local populations through community-based management models, which involve them in decision-making, monitoring, and enforcement. This will help create a sense of ownership and increase the sustainability of conservation efforts. Finally, the integration of mangrove conservation into broader climate change adaptation strategies is essential. The Red Sea countries should work toward aligning mangrove conservation efforts with national and regional climate change policies, recognizing mangroves' role in carbon sequestration and climate resilience.

2.7 Conclusion

The study has shed light on the significant role mangrove ecosystems play in maintaining coastal resilience, while underscoring the critical importance of community involvement in conservation efforts. Through the case studies and observations from the Red Sea region, it becomes clear that mangroves serve as vital buffers against storm surges, coastal erosion, and the impacts of climate change. They also contribute to biodiversity, support fisheries, and provide local communities with resources for livelihood. The lessons from the region emphasize the necessity of a multifaceted approach to mangrove conservation, combining environmental science with local knowledge and engagement. A key lesson from the Red Sea region is the understanding that mangrove ecosystems are under increasing threat due to human activities, such as coastal development, pollution, and unsustainable resource extraction. However, there are promising examples of effective community-led conservation projects, where local stewardship has helped to restore and protect mangrove forests. These efforts highlight the value of integrating traditional knowledge with scientific management strategies, ensuring that conservation efforts are culturally relevant and sustainable. The success stories suggest that involving local communities in monitoring, managing, and benefiting from these ecosystems can lead to better long-term outcomes for both the environment and the people who depend on these habitats. Looking ahead, the future of mangrove conservation in the Red Sea region hinges on the continuation and expansion of these community-based efforts. Strengthening policies that support mangrove protection, along with improved monitoring and research, will be essential. Moreover, there is a need for greater integration of mangrove conservation into national and regional environmental policies, ensuring that these ecosystems are not only protected but also sustainably managed for future generations. By addressing the social, economic, and environmental dimensions of mangrove conservation, it is possible to create a resilient coastal ecosystem that benefits both nature and the communities that rely on it. Ultimately, integrating local communities with environmental policies is key to building long-term resilience along the Red Sea coast. Collaborative approaches that prioritize both human and ecological well-being will foster a sustainable future where mangroves thrive, ensuring the continued protection of coastal ecosystems and the livelihoods they support.

References

Abdel-Hamid, A., Dubovyk, O., Abou El-Magd, I., & Menz, G. (2018). Mapping mangroves extents on the Red Sea coastline in Egypt using polarimetric SAR and high resolution optical remote sensing data. *Sustainability, 10*(3), 646.

Abou Samra, R. M., Selim, E. M. M., Hefny, W. A., & El-Gammal, M. E. Biomass carbon stock and carbon sequestration potentiality in mangrove ecosystem along the Egyptian Red Sea Coast. *Scientific Journal for Damietta Faculty of Science, 14*(1), 41–54. https://sjdfs.journals.ekb.eg/article_3 48068_619abb06884a3254962b1d4238a24c7f.pdf

Afefe, A. (2021). Linking territorial and coastal planning: Conservation status and management of mangrove ecosystem at the Egyptian-African Red Sea coast. *Aswan University Journal of Environmental Studies, 2*(2), 91–114.

Afefe, A. A., Khedr, A. H. A., Abbas, M. S., & Soliman, A. S. (2021). Responses and tolerance mechanisms of mangrove trees to the ambient salinity along the Egyptian Red Sea Coast. *Limnological Review, 21*(1), 3–13.

Afefe, A.A., S Abbas, M., Sh Soliman, A., A Khedr, A. H., & E Hatab, E. B. (2019). Physical and chemical characteristics of mangrove soil under marine influence. A case study on the Mangrove Forests at Egyptian-African Red Sea Coast. *Egyptian Journal of Aquatic Biology and Fisheries, 23*(3), 385–399.

Ahmed, E. A., & Abdel-Hamid, K. A. (2007). Zonation pattern of Avicennia marina and Rhizophora mucronata along the Red Sea Coast, Eglypt. *World Applied Sciences Journal, 2*(4), 283–288.

Alhassan, A. B., & Aljahdali, M. O. (2021). Nutrient and physicochemical properties as potential causes of stress in mangroves of the central Red Sea. *PLoS ONE, 16*(12), Article e0261620.

Allen, J. A., & Duke, N. C. (2006). *Bruguiera gymnorrhiza (large-leafed mangrove) Rhizophoraceae (mangrove family)*. https://www.researchgate.net/publication/240618767_Bruguiera_gymnorrhiza_large-leafed_mangrove

Awad, A., El-Sammak, A., Elshazly, A., & El-Masry, E. A. (2023). Carbon Sequestration in mangrove sediments as climate change mitigation tool: A case study from the Red Sea, Egypt. *Egyptian Journal of Aquatic Biology & Fisheries, 27*(4).

Awimbo, J., Barrow, E., & Karaba, M. (2004). Community-based natural resource management in the IGAD region. *IUCN-EARO & IGAD*, 249p.

Beyer, J., Staalstrøm, A., Wathne, B. M., Omer, R. K., & Ahmed, S. E. (2015). *Marine ecological baselines and environmental impact assessment studies in the Sudanese coastal zone. A review.* http://hdl.handle.net/11250/2365778

Blanco-Sacristán, J., Johansen, K., Duarte, C. M., Daffonchio, D., Hoteit, I., & McCabe, M. F. (2022). Mangrove distribution and afforestation potential in the Red Sea. *Science of the Total Environment, 843*, Article 157098.

CBD (2019). *Eritrea: 6th National report to the convention on biological diversity*, 134p. https://www.cbd.int/doc/nr/nr-06/er-nr-06-en.pdf

Cziesielski, M. J., Duarte, C. M., Aalismail, N., Al-Hafedh, Y., Anton, A., Baalkhuyur, F., Aranda, M., et al. (2021). Investing in blue natural capital to secure a future for the Red Sea ecosystems. *Frontiers in Marine Science, 7*, Article 603722.

Dicks, B. (1986). Oil and the black mangrove, Avicennia marina in the northern Red Sea. *Marine Pollution Bulletin, 17*(11), 500–503.

Diop, E. S., Gordon, C., Semesi, A. K., Soumaré, A., Diallo, N., Guissé, A., & Ayivor, J. S., et al. (2002). Mangroves of Africa. In *Mangrove ecosystems: Function and management* (pp. 63–121). Springer Berlin Heidelberg.

Elasha, B. O. (2010). Mapping of climate change threats and human development impacts in the Arab region. *UNDP Arab Development Report–Research Paper Series, UNDP Regiona Bureau for the Arab States.* https://www.undp.org/sites/g/files/zskgke326/files/migration/arabstates/Mapping-of-Climate-Change-Threats.pdf

Elmogy, S. A. (2024). Blue carbon's importance in preserving sustainable growth in the Red Sea in the face of climate change. *The International Journal of Tourism and Hospitality Studies, 6*(2), 21–46.

Gebreyohanns, M. Z. (2006). *The state of tourism in Eritrea: Tourism development in the Dahlak Islands.* University of Pretoria (South Africa). https://www.proquest.com/openview/b50bf2d8d5bd44453e3dce03574fac28/1?pq-origsite=gscholar&cbl=2026366&diss=y

Genevier, L. G., Jamil, T., Raitsos, D. E., Krokos, G., & Hoteit, I. (2019). Marine heatwaves reveal coral reef zones susceptible to bleaching in the Red Sea. *Global Change Biology, 25*(7), 2338–2351.

Gladstone, W., Tawfiq, N., Nasr, D., Andersen, I., Cheung, C., Drammeh, H., Lintner, S., et al. (1999). Sustainable use of renewable resources and conservation in the Red Sea and Gulf of Aden: Issues, needs and strategic actions. *Ocean & Coastal Management, 42*(8), 671–697.

Haile, A. (1999). *Inter-sectoral co-ordination and the legal framework to protect the Eritrean marine environment.* World Maritime University, 101p. https://commons.wmu.se/cgi/viewcontent.cgi?article=1235&context=all_dissertations

Hariri, K. I., Nichols, P., Krupp, F., Mishrigi, S., Barrania, A., Ali, A. F., & Kedidi, S. M. (2000). *Status of the living marine resources in the Red Sea and Gulf of Aden region and their management.* Strategic Action Programme for the red Sea and Gulf of Aden, final Report, 1–148.

Haseeba, K. P., Aboobacker, V. M., Vethamony, P., & Al-Khayat, J. A. (2025). Significance of Avicennia Marina in the Arabian Gulf Environment: A review. *Wetlands, 45*(1), 1–27.

Khalil, A. S. (2015). Mangroves of the red sea. In *The Red Sea: The formation, morphology, oceanography and environment of a young ocean basin* (pp. 585–597).

Kitula, R. (2022). *Sustainability of land management approaches and practices applied in Eastern African forests*. African Forest Forum (AFF). AFF Working Paper. African Forest Forum, Nairobi. https://afforum.org/oldaff/sites/default/files/English/English_243.pdf

Lakhouit, A. (2019). Tourism impact on marine ecosystems in the north of Red Sea. *Journal of Sustainable Development, 13*(1), 10–17.

Little, D. I. (2018). Mangrove restoration and mitigation after oil spills and development projects in East Africa and the Middle East. *Threats to Mangrove Forests: Hazards, Vulnerability, and Management*, 637–698.

Madkour, H. A. (2015). *Detection of damaged areas due to tourism development along the Egyptian Red Sea coast using GIS, remote sensing and foraminifera*. State of the Art, National Institute of Oceanography and Fisheries, Red Sea Branch, 135p.

Madkour, H. A., & Mohammed, A. W. (2008). Nature and geochemistry of surface sediments of the mangrove environment along the Egyptian Red Sea coast. *Environmental Geology, 54*(2), 257–267.

Manohar, S. M., Yadav, U. M., Kulkarnii, C. P., & Patil, R. C. (2023). *An overview of the phytochemical and pharmacological profile of the spurred mangrove Ceriops tagal (Perr.) CB Rob*.

Mehanzel, S. H. (2012). *Impact of goat development project on livelihood assets: The case of Northern Red Sea region in eritrea*. Master's Degree in Management of Development,(Leeuwarden: Van Hall Larenstein University of Applied Sciences, 2012).

Moussalli, E., & Feidi, I. H. (2010). Red Sea and gulfs fisheries. *Handbook of Marine Fisheries Conservation and Management, 1*, 426.

Moustafa, A. A., Abdelfath, A., Arnous, M. O., Afifi, A. M., Guerriero, G., & Green, D. R. (2023). Monitoring temporal changes in coastal mangroves to understand the impacts of climate change: Red Sea, Egypt. *Journal of Coastal Conservation, 27*(5), 37.

Naidoo, G. (2023). The mangroves of Africa: A review. *Marine Pollution Bulletin, 190*, 114859.

Nasr, D. H., Hamza, M. E., Ali, M. E., & Hamad, A. E. (2015). Management and conservation of marine biodiversity in Sudan. *Red Sea University Journal*, 2(June 2012). https://citeseerx.ist.psu.edu/document?repid=rep1&type=pdf&doi=cd589dbe0e87bf60120a576431018505306584c8

Ogbazghi, W. (2018). Agro-climatic and environmental hazards and mitigation measures for the Northern and Southern Red Sea zone administrations of Eritrea. *J. Eritrean Stud, 8*, 85–131.

Osman, E. O., Smith, D. J., Ziegler, M., Kürten, B., Conrad, C., El-Haddad, K. M., Suggett, D. J., et al. (2018). Thermal refugia against coral bleaching throughout the northern Red Sea. *Global Change Biology, 24*(2), e474–e484.

Paruğ, Ş, Jabbr, A., & Lazrag, F. A. (2024). Enhancing sustainable fisheries and aquaculture in North Africa: Challenges, opportunities, and future directions. *Menba Kastamonu Üniversitesi Su Ürünleri Fakültesi Dergisi, 10*(1), 90–105.

PERSGA. (2024). *Regional organization for the conservation of the environment of the Red Sea and Gulf of Aden*. https://icriforum.org/members/regional-organization-for-the-conservation-of-the-environment-of-the-red-sea-and-gulf-of-aden-persga/

Pilcher, N., & Alsuhaibany, A. (2020). *Regional status of coral reefs in The Red Sea and The Gulf of Aden-Middle East-2000*. https://archive.iwlearn.org/persga.org/Documents/Coral_reef_status_RSGA_2000_1_.pdf

Sarhan, M., & M Elmahdy, Y. (2023). The socio-economic impact of the COVID-19 pandemic on ecotourism: A case study of Wadi El Gemal national park in Egypt. In *Towards sustainable and resilient tourism futures* (pp. 189–203). Erich Schmidt Verlag GmbH & Co. KG.

Sector, N. C. (2006). *Protected areas of Egypt: Towards the future* (p. 71). Egyptian Environmental Affairs Agency.

Shaalan, I. M. (2005). Sustainable tourism development in the Red Sea of Egypt threats and opportunities. *Journal of Cleaner Production, 13*(2), 83–87.

Shaltout, K. H., Khalaf-Allah, A., & El-Bana, M. (2005). Environmental characteristics of the mangrove sites along the Egyptian Red Sea coast. *Assessment and management of mangrove forests in Egypt for sustainable utilization and development: A project funded by ITTO (Japan) and supervised by MALR/MSEA—EEAA*.

Siddig, A. A., Magid, T. D. A., EL-Nasry, H. M., Hano, A. I., & Mohammed, A. A. (2018). Biodiversity in Sudan. In *Global biodiversity* (pp. 273–294). Apple Academic Press.

Sugga, A. A., Hummed, M. S., Ahmed, M. F., & Ahmed, A. A. (2024). Assessment of climate change impacts on blue economy resources in Sudan: A case study of maritime shipping. *Handbook of Sustainable Blue Economy*, 1–24.

Tesfamichael, D. (2012). *Assessment of the Red Sea ecosystem with emphasis on fisheries* (Doctoral dissertation, University of British Columbia). https://open.library.ubc.ca/media/stream/pdf/24/1.0072911/2

USAID. (2012). *Sudan environmental threats and opportunities assessment with special focus on biological diversity and tropical forest*, 66p. https://citeseerx.ist.psu.edu/document?repid=rep1&type=pdf&doi=a8973b3ed19d91ed0e861aaf5fb0e5d7edd560b8

Waleed, T. A., Abdel-Maksoud, Y. K., Kanwar, R. S., & Sewilam, H. (2024). Mangroves in Egypt and the Middle East: Current status, threats, and opportunities. *International Journal of Environmental Science and Technology*, 1–38.

Youssef, N. A. E., Tonbol, K., Hassaan, M. A., Mandour, A., El-Sikaily, A., Elshazly, A., & Shabaka, S. (2024). Blue carbon assessment in Avicennia marina sediments and vegetation along the Red Sea Coast of Egypt: Improving methods and insights. *Continental Shelf Research, 280*, Article 105299.

3

Central Africa's Coast: Mangrove Ecosystems and Community Resilience

Abstract Central Africa's coastal zones, notably in Cameroon and Gabon, are home to significant mangrove ecosystems that provide vital ecological and socio-economic services. Mangroves in this region are essential for coastal protection, biodiversity support, and the sustenance of local fisheries, which are crucial for the livelihoods of coastal communities. However, industrialization, deforestation, and the expansion of oil exploration and coastal development pose serious threats to these vital ecosystems. Despite these challenges, local communities have shown resilience by adopting traditional ecological knowledge and engaging in community-led mangrove conservation initiatives. These efforts highlight the potential for integrating indigenous practices with modern conservation techniques to ensure sustainable mangrove management. This study explores the biodiversity of Central African mangroves, focusing on species such as *Rhizophora spp.*, *Avicennia spp.*, *Laguncularia racemosa*, and *Bruguiera spp.*, and their role in carbon sequestration and environmental health. Additionally, it examines case studies of community-driven mangrove restoration projects in Cameroon and Gabon, demonstrating the effectiveness of grassroots involvement in conservation efforts. The study emphasizes the importance of strengthening community resilience through ecosystem-based adaptation and providing policy recommendations for the sustainable management of mangrove forests. Ultimately, this research underscores the necessity of multi-stakeholder approaches, involving governments, NGOs, and local communities, to safeguard the future of Central Africa's mangrove ecosystems and ensure their long-term resilience against environmental and socio-economic challenges.

Keywords Central Africa · Mangrove ecosystems · Community resilience · Biodiversity · Sustainable practices · Conservation initiatives

3.1 Introduction to Central Africa's Coastal Zones

The coastal zones of Central Africa, particularly in countries like Cameroon, Equatorial Guinea, and Gabon, represent crucial ecological and socio-economic regions. These coastal areas are characterized by an intricate blend of diverse ecosystems, including estuaries, beaches, and vital mangrove forests (Diop et al., 2014). This region plays a pivotal role in the ecological balance, local economies, and the cultural identity of its communities. The importance of these coastal zones is highlighted not only by their rich biodiversity but also by their economic significance, as they sustain fisheries, agriculture, and tourism. Cameroon's coastal zone stretches over 400 kilometers along the Gulf of Guinea, encompassing both sandy beaches and extensive mangrove swamps (Ngoran, 2014). Similarly, Gabon's coastal region, which is approximately 800 kilometers long, features similar ecosystems, with vast mangrove forests, especially along its southern coastline. The mangroves in these regions are particularly significant due to their role in coastal protection, biodiversity conservation, and carbon sequestration (Ajonina et al., 2014; Feka & Morrison, 2017; Kauffman & Bhomia, 2017). These areas are home to numerous species of fish, shellfish, and crustaceans, which support local fishing industries. Furthermore, the rich biodiversity of the mangroves in Cameroon, Equatorial Guinea, Gabon, and other countries in Central Africa has made them essential for migratory birds, providing nesting and feeding grounds for various avian species.

In Central Africa, the ecological and socio-economic importance of mangrove forests cannot be overstated. Mangroves serve as natural buffers, protecting coastal communities from storm surges and erosion, which are becoming more frequent due to climate change (Corcoran et al., 2007). Their dense root systems stabilize sediment, reduce wave impact, and prevent shoreline erosion. Moreover, mangroves contribute to the health of marine ecosystems by acting as nurseries for many marine species, including commercially important fish and crustaceans (Ajonina, 2022). This biodiversity is vital for the livelihoods of local communities who depend on fisheries for food and income. The extraction of resources from mangrove forests, such as wood and charcoal, also plays a significant role in local economies. However, these coastal zones are increasingly under threat due to rapid industrialization and deforestation. The expansion of oil, gas, and timber industries in Cameroon, Equatorial Guinea, and Gabon has led to significant environmental degradation. Oil spills, industrial waste, and illegal logging have damaged the mangrove ecosystems, reducing their ability to provide ecological services such as carbon sequestration and coastal protection. Equally, the conversion of mangrove land for agriculture and urban development has further intensified the loss of these critical habitats (Asangwe, 2006; Feka & Ajonina, 2011; Lontsi et al., 2023; Munji et al., 2013). Local communities, who have long depended on the resources provided by mangroves, are facing severe challenges, including reduced fish stocks, loss of livelihood, and increased vulnerability to climate-induced natural disasters.

3.2 Rich Biodiversity and Mangrove Ecosystem Services Along the Coasts of Central Africa

The mangrove ecosystems along the coasts of Central Africa (Fig. 3.1) are among the most biologically rich and ecologically important habitats in the region (Polidoro et al., 2017). These coastal forests, which are characterized by a unique assemblage of salt-tolerant trees and shrubs, thrive in the intertidal zones where freshwater meets seawater. The mangrove forests of Central Africa, which extend across countries like Gabon, Cameroon, and Equatorial Guinea, harbor diverse species of flora and fauna, making them critical to both local biodiversity and the global environment.

Central African mangrove forests are home to a wide array of plant and animal species adapted to the harsh, saline conditions of coastal environments. The primary tree species include *Rhizophora*, *Avicennia*, *Laguncularia*, and *Sonneratia* (Table 3.1), each contributing to the structural complexity of the ecosystem (Jalloh et al., 2012). These forests provide habitat for a range of species, including birds, fish, invertebrates, and marine mammals. The mangrove roots serve as nurseries for juvenile fish species, which later migrate into open waters, making these forests crucial for maintaining both local and regional biodiversity (Carrere, 2009). Furthermore, the presence of mangroves supports the survival of threatened species such as the West African manatee

Fig. 3.1 Map showing distribution of mangroves along the coasts of Central Africa

(*Trichechus senegalensis*) and various species of sea turtles (Laudisoit et al., 2017).

Mangroves are vital to the livelihoods of millions of people living in the coastal regions of Central Africa. These ecosystems provide numerous direct benefits to local communities, particularly in the form of fishery resources. The mangrove roots create sheltered environments that offer breeding grounds for a variety of commercially important fish species, including barracuda, snapper, and shrimp (Ajonina & Eyango, 2020). Local fishers rely on these species for both subsistence and income generation. Furthermore, mangrove ecosystems support artisanal fisheries, which provide employment to coastal populations. In addition to their support for fisheries, mangroves also contribute to coastal protection. The dense root systems of mangrove trees reduce wave energy and help prevent coastal erosion, a crucial service in the face of rising sea levels and storms associated with climate change (Nwabueze, 2024; Santos et al., 2024). As such, mangroves act as natural barriers, safeguarding coastal communities from flooding and storm surges, which directly benefits both human settlements and local economies.

Mangroves play an essential role in mitigating climate change by acting as significant carbon sinks. The dense vegetation and waterlogged soil conditions in mangrove forests make them highly effective at capturing and storing carbon dioxide from the atmosphere (Ajonina et al., 2015; Bumtu et al., 2020; Gaë et al., 2024). These forests sequester carbon at a rate several times higher than tropical rainforests, making them one of the most efficient carbon storage systems on Earth. Mangrove ecosystems along the coasts of Central Africa contribute substantially to global efforts to combat climate change through carbon sequestration. Beyond carbon storage, mangroves also contribute to environmental health by improving water quality. Their intricate root systems filter out pollutants and trap sediments, preventing the degradation of coral reefs and other coastal ecosystems. Mangroves also support nutrient cycling, enhancing the productivity of both marine and terrestrial ecosystems.

Mangrove species in Central Africa play essential roles in coastal protection and biodiversity (Table 3.1). *Rhizophora* spp., also known as red mangroves, are recognized for their distinctive prop roots, which

Table 3.1 Mangrove species in Central Africa

Mangrove species	Description	References
Rhizophora spp.	Species like *Rhizophora mangle*, commonly known as red mangroves, are recognized for their unique prop roots that offer support and stability. These trees play a vital role in safeguarding coastal areas and act as essential nurseries for marine organisms	Hou (1955), Saenger and Bellan (1995), Diop et al. (2002) Corcoran et al. (2007), Ajonina (2008), Nfotabong-Atheull et al. (2011), Kauffman and Bhomia (2017), and Ngeve et al. (2021)
Avicennia spp.	Species like *Avicennia germinans*, commonly referred to as black mangroves, are identifiable by their aerial roots and salt-excreting glands. These trees are highly adapted to thrive in saline environments and play a crucial role in shaping the structure of mangrove ecosystems	Saenger and Bellan (1995), Diop et al. (2002), Corcoran et al. (2007), Ajonina (2008), and Kauffman and Bhomia (2017)
Laguncularia racemosa	*Laguncularia racemosa*, commonly known as white mangrove, features a less prominent root system compared to red and black mangroves. Typically found in the upper intertidal zones, this species plays an important role in maintaining the biodiversity of mangrove ecosystems	Lonard Robert et al. (2020) and Sereneski-Lima et al. (2021)
Bruguiera spp.	*Bruguiera gymnorhiza* and similar species are recognized for their unique knee-like roots and their function in sediment stabilization. These trees are well-suited to fluctuating salinity levels and play a key role in enhancing the ecological diversity of mangrove forests	Diop et al. (2002) and Corcoran et al. (2007)

offer stability and serve as nurseries for marine life. These species are vital for coastal protection (Ajonina, 2008; Corcoran et al., 2007; Diop et al., 2002; Hou, 1955; Kauffman & Bhomia, 2017; Nfotabong-Atheull et al., 2011; Ngeve et al., 2021; Saenger & Bellan, 1995). *Avicennia* spp., or black mangroves, are adapted to high salinity with aerial roots and salt-excreting glands, contributing significantly to mangrove forest structure (Ajonina, 2008; Corcoran et al., 2007; Diop et al., 2002; Kauffman & Bhomia, 2017; Saenger & Bellan, 1995). *Laguncularia racemosa*, known as white mangrove, features a less pronounced root system and is commonly found in upper intertidal zones. This species is vital for maintaining biodiversity in mangrove ecosystems (Lonard Robert et al., 2020; Sereneski-Lima et al., 2021). Finally, *Bruguiera* spp., such as *Bruguiera gymnorhiza*, are notable for their knee-like roots, which help stabilize sediment. These trees thrive in varying salinity levels and contribute to the ecological complexity of mangrove forests (Corcoran et al., 2007; Diop et al., 2002). These species collectively play crucial roles in ecosystem stability and biodiversity.

3.3 Traditional Knowledge and Sustainable Practices for Mangrove Ecosystems in Central Africa

Mangrove ecosystems in Central Africa are crucial for coastal protection, biodiversity conservation, and the livelihoods of local communities. The region's mangrove forests, spanning countries like Cameroon, Gabon, Republic of Congo, Democratic Republic of Congo, and Equatorial Guinea, have long been managed through a combination of indigenous knowledge and practices. These traditional methods have not only ensured the sustainability of these ecosystems but also supported the local economies. However, the integration of modern conservation techniques with traditional knowledge is essential to address the contemporary threats facing mangrove ecosystems.

Indigenous communities in Central Africa have a profound relationship with mangrove forests (Nfotabong-Atheull et al., 2011). The knowledge passed down through generations has been fundamental in maintaining ecological balance and sustainable resource use. One key traditional practice is the selective harvesting of mangrove species. For instance, certain mangrove species are harvested only at specific times of the year to prevent overexploitation, a practice that aligns with the concept of sustainable resource management (Din et al., 2008). Villages typically maintain taboos and customs that regulate the exploitation of mangrove resources, including restrictions on cutting certain trees during particular seasons, or the prohibition of mangrove harvesting in specific areas to allow for regeneration. Local communities also use mangroves for a variety of purposes, such as fishing, medicinal plants, construction materials, and fuel. The method of "shifting exploitation," where communities move to different mangrove stands to prevent overuse of a single area, is another form of sustainability practiced in some Central African coastal regions. According to Din et al. (2008), this rotational resource management is a strategic form of sustainability that ensures the recovery of mangrove stands and prevents soil degradation.

The integration of traditional ecological knowledge (TEK) with modern conservation techniques holds significant promise for enhancing mangrove conservation in Central Africa. Research by Munji et al. (2014) underscores the importance of local ecological knowledge in understanding the intricate relationships within mangrove ecosystems. TEK, such as knowledge about seasonal patterns, the health of mangrove species, and the relationships between coastal tides and plant growth, can complement scientific research and inform more effective conservation strategies. For example, community-based management schemes that incorporate indigenous knowledge have been shown to promote sustainable mangrove management. In Cameroon, for example, community-led conservation initiatives are increasingly being linked to formal government policies and international conservation frameworks, such as the Ramsar Convention on Wetlands. These efforts merge local ecological wisdom with modern scientific tools like Geographic Information Systems (GIS) for mapping mangrove areas and monitoring their health, as noted by Fonteh et al. (2009), Ellison and Zouh (2012), and Findi and Wantim (2022). This integration helps strengthen local stewardship and raises awareness about the value of mangrove ecosystems.

Local perceptions of mangrove conservation and sustainable use vary across Central Africa but are generally influenced by traditional reliance on these ecosystems. In some areas, mangroves are viewed as essential for the survival of local communities, particularly for food security, due to their role in supporting fish populations. However, growing pressures from urbanization, industrial development, and climate change have led to diminishing mangrove resources, prompting concerns about the future of these ecosystems. Communities often perceive the loss of mangroves as detrimental not only to their livelihoods but also to cultural practices tied to the forests. Thus, conservation efforts that acknowledge and incorporate these perceptions can improve community support. The involvement of local people in decision-making processes, through participatory conservation programs, has been shown to yield more successful outcomes, as local stakeholders have a vested interest in the long-term health of mangrove ecosystems.

3.4 Community-Led Initiatives in Mangrove Conservation in Central Africa

Mangrove ecosystems, crucial for biodiversity, coastal protection, and carbon sequestration, are rapidly degrading due to human activities such as urbanization, deforestation, and industrial development. Central Africa, with its rich mangrove forests spanning countries like the Republic of Congo, Cameroon, Equatorial Guinea, Democratic Republic of Congo, and Gabon, faces significant threats to these valuable ecosystems. However, community-led initiatives have emerged as a powerful tool for restoration and conservation in these regions, offering sustainable solutions rooted in local knowledge and practices. In Cameroon, coastal mangroves are under threat from logging, shrimp farming, and land reclamation (Ajonina, 2022; Din et al., 2017; Feka & Manzano, 2008; Moudingo et al., 2020). A noteworthy community-led project is the one spearheaded by the Cameroon Mangroves and Wetlands Network (CMN) in the coastal towns of Limbe and Douala. This initiative focuses on both the protection and rehabilitation of mangroves through active community engagement and sustainable resource management. Local fishermen and women, who heavily rely on mangroves for livelihoods, have become key participants in these efforts. The CMN works with local communities to restore degraded mangrove areas, monitor the health of these ecosystems, and promote sustainable fishing practices that do not harm the mangroves (CMN, 2024). In Equatorial Guinea, mangrove conservation efforts have been similarly driven by local communities, with a strong emphasis on community education. In the region of Bioko Island, local organizations such as the Bioko Biodiversity Protection Program (BBPP) have been involved in mangrove restoration programs (BBPP, 2024). These initiatives aim to rebuild mangrove forests that have been degraded due to industrial activity, including oil extraction. The program incorporates educational campaigns to raise awareness about the ecological and economic benefits of mangroves, thereby empowering local communities to protect and sustainably manage their coastal resources (BBPP, 2024). In Gabon, the government has collaborated with local communities in a series of successful mangrove restoration projects. The Gabonese National Agency for National Parks has worked with coastal communities, especially those living near the Pongara National Park, to plant mangrove seedlings and prevent illegal logging. In addition to restoration activities, community participation is critical in monitoring the health of the mangroves. These projects, which focus on collaboration with local authorities and NGOs, have fostered a sense of ownership among community members, who understand the importance of these ecosystems for their livelihoods (ANPN, 2024; Olam, 2024).

Central to the success of community-led mangrove conservation in these countries is the emphasis on education and participation. In Cameroon, community engagement programs have been implemented to train locals in sustainable land-use practices, such as agroforestry and alternative livelihoods that reduce pressure on mangroves. Educational initiatives are essential for helping community members understand the benefits of mangrove forests, including their role in coastal protection, carbon sequestration, and biodiversity conservation. Similarly, in Equatorial Guinea, community involvement in mangrove restoration is enhanced through local schools and workshops. Programs that focus on environmental stewardship and the sustainable use of natural resources provide community members, especially youth, with the tools to participate in restoration activities and spread knowledge within their communities. In Gabon, the success of mangrove restoration is rooted in the active involvement of local communities in every stage of the process. From planting mangrove seedlings to monitoring ecosystem health, communities play a vital role. This involvement empowers them to take ownership of the conservation effort, ensuring that restoration projects are not only sustainable but also culturally relevant.

Several community-driven mangrove rehabilitation projects in Central Africa have demonstrated remarkable success. The rehabilitation of mangroves in Limbe and Douala, Cameroon, has not only led to the restoration of vital coastal ecosystems but has also brought tangible benefits to local communities. These include improved fish stocks and increased resilience to coastal flooding (Ajonina et al., 2016; Boubakary et al., 2019, 2024; Fopi et al., 2021; Jean-Hude et al., 2016; Mumbang et al., 2025; Quenum et al., 2024; Shokoleu et al., 2024). In Equatorial Guinea, the restoration efforts on Bioko Island have led to the rejuvenation of previously degraded mangrove forests, providing vital ecosystem services, including improved water quality and enhanced biodiversity (BBPP, 2024). These successes have garnered attention from local governments and international organizations, further strengthening the commitment to mangrove conservation. In Gabon, the restoration projects in Pongara National Park have seen a resurgence in mangrove biodiversity, along with increased awareness of the importance of preserving coastal ecosystems. The active participation of local communities has played a central role in these outcomes, making the projects both successful and sustainable (Ajonina et al., 2016).

3.5 Industrialization, Deforestation, and the Impact on Mangroves in Central Africa

Industrialization has accelerated economic growth across Central Africa, but it has also come at the expense of vital ecosystems, particularly mangroves. Mangroves, which thrive along tropical coastlines, are invaluable for their role in protecting coastal communities, supporting biodiversity, and sequestering carbon. However, activities like logging, oil exploration, and coastal development have significantly disrupted these habitats in Central Africa. The most direct threat to mangrove ecosystems in Central Africa arises from logging. Mangroves are heavily exploited for timber and fuelwood, particularly for construction and charcoal production. This has been particularly prevalent in countries like Cameroon, Gabon, Republic of Congo, Democratic Republic of Congo, and Equatorial Guinea, where the demand for forest products is high. According to Feka and Manzano (2008), Feka and Morrison (2017), and Ajonina (2022), unsustainable logging practices have led to widespread deforestation of mangrove forests, diminishing their capacity to perform essential ecological services. Oil exploration has also had a profound impact on mangrove ecosystems. Oil spills, pipeline leakages, and the construction of oil platforms disrupt not only the mangrove habitats but also the delicate balance of the surrounding marine environments. In the Gulf of Guinea, for instance, oil spills have contaminated large swathes of coastal ecosystems, decimating mangrove populations and harming the species that depend on them (McGlade et al., 2002; Okafor-Yarwood, 2020; Onyena & Sam, 2020). The introduction of toxic substances and the disruption of hydrodynamic patterns further exacerbate the loss of biodiversity. Coastal development, such as the construction of ports, resorts, and infrastructure, is another major driver of mangrove degradation. As urbanization expands along Central Africa's coastline, large areas of mangrove forests are cleared to make way for development projects. This is particularly evident in rapidly growing urban centers like Libreville, Douala, and Malabo, where development pressures threaten nearby mangrove areas (Bissonnette et al., 2024; Chretien et al., 2009; Din et al., 2017; Nfotabong-Atheull et al., 2011, 2013; Ngoran et al., 2015). Such encroachment leads to habitat fragmentation, making it harder for mangroves to recover.

Several strategies have been proposed to mitigate the impacts of industrialization on mangroves. First, promoting sustainable logging practices is crucial. The certification of mangrove timber, along with establishing sustainable harvest limits, can reduce the pressure on these ecosystems. The involvement of local communities in sustainable resource management is key, as community-based forestry initiatives have been shown to successfully preserve mangrove forests (Forkam et al., 2020). Restoration efforts also play an important role in protecting mangroves. Several countries in Central Africa have launched mangrove restoration programs, which focus on replanting mangrove species in degraded areas and enhancing natural regeneration. The use of environmental impact assessments (EIAs) for development projects can ensure that the health of mangrove habitats is considered before major infrastructure projects proceed. Finally, promoting alternative livelihoods for coastal communities, such as ecotourism or sustainable aquaculture, can reduce reliance on mangrove ecosystems for economic gain. Integrating environmental education into local development agendas can raise awareness about the importance of mangroves and encourage more sustainable practices.

In response to the mounting threats to mangroves, several countries in Central Africa have put in place legal frameworks to protect these vital ecosystems. International agreements such as the 1971 Ramsar Convention on Wetlands and the 1992 United Nations Convention on Biological Diversity provide a foundation for regional and national legal frameworks aimed at conserving mangroves. In Central Africa, countries like Equatorial Guinea, Republic of Congo, Cameroon, Democratic Republic of Congo, and Gabon have ratified these agreements, enacting national laws to regulate mangrove destruction. At the national level, legislation such as Cameroon's 1994 Forestry Law (updated to the 2024 Forestry Law) and the 1996 Environmental Law provides mechanisms for managing forest resources and protecting coastal ecosystems. Additionally, the creation of protected areas, such as national parks and wildlife reserves (like the Douala-Edea National Park in Cameroon and the Pongara National Park in Gabon), ensures that significant mangrove ecosystems are preserved for future generations. However, enforcement of these laws remains a challenge due to limited resources and weak institutional frameworks.

3.6 Resilience Building Through Mangroves and Policy Implications in Central Africa

Mangroves, often referred to as "coastal guardians," are vital ecosystems that provide essential services such as protecting shorelines from erosion, mitigating storm surges, enhancing biodiversity, and supporting the livelihoods of local communities. In Central Africa, mangrove forests along the coasts of Gabon, Republic of Congo, Cameroon, Democratic Republic of Congo, and Equatorial Guinea, face growing threats from deforestation, climate change, and unsustainable exploitation. Resilience building through ecosystem-based adaptation (EbA) offers a promising strategy to not only safeguard these unique ecosystems but also enhance the resilience of communities that depend on them.

Ecosystem-based adaptation is an approach that leverages natural ecosystems and their services to reduce vulnerability to climate change impacts. In Central Africa, where many coastal communities depend on mangroves for food, fuelwood, and income, EbA can be an effective strategy for resilience building. Mangroves protect against sea-level rise and storm surges, vital for safeguarding communities living in low-lying coastal areas. According to Din et al. (2016), Evariste et al. (2018), and Fongnzossie et al. (2022), EbA strengthens the resilience of vulnerable communities by integrating ecosystem services into adaptation planning. Additionally, mangroves are critical carbon sinks, thus playing a dual role in both adaptation and mitigation of climate change (Awazi et al., 2024). Through the restoration and preservation of mangrove forests, communities can reduce their exposure to the negative effects of climate change while benefiting from enhanced fishery resources and improved water quality. For instance, in the coastal regions of Gabon and Cameroon, where communities rely heavily on fishing, the restoration of degraded mangrove areas can help restore fish populations and sustain local economies. Furthermore, EbA strategies, which emphasize local knowledge and participation, can empower communities to take ownership of their environment, fostering a sense of stewardship and sustainability.

Effective mangrove management in Central Africa requires policies that balance conservation with community needs. Several policy recommendations are essential to ensure the sustainable management of these vital ecosystems including inclusive policy frameworks, enhancing knowledge and capacity, legal protection and enforcement, and incentivizing sustainable livelihoods. Governments should integrate mangrove conservation into national climate change adaptation and biodiversity strategies, ensuring that local communities are active participants. Policies should encourage community-led restoration efforts and provide financial incentives for sustainable mangrove use (Ajonina, 2007; Brunnschweiler & Luisetti, 2022). Governments and NGOs should invest in capacity-building programs that educate local populations about the importance of mangroves and how to sustainably manage them. Research and data collection should also be prioritized to understand the full ecological and economic value of mangrove ecosystems. Strengthening the legal frameworks surrounding mangrove conservation and ensuring robust enforcement mechanisms is essential. Policies should address illegal logging, land conversion for agriculture, and overexploitation of mangrove resources. Regular monitoring and compliance checks should be conducted to prevent illegal activities. Alternative livelihood programs, such as sustainable aquaculture and ecotourism, should be promoted to reduce the dependence on destructive practices like unsustainable fishing or mangrove timber extraction.

Governments play a crucial role in supporting mangrove conservation through policy development, resource allocation, and international cooperation. The Central African governments must prioritize the preservation of mangrove ecosystems by aligning their policies with global environmental agreements such as the Paris Agreement and the Convention on Biological Diversity. Furthermore, collaborative efforts with NGOs can enhance the scope and effectiveness of conservation initiatives. Non-governmental organizations (NGOs) also have a pivotal role in implementing on-the-ground conservation projects. They can facilitate community engagement, raise awareness about the importance of mangrove ecosystems, and help secure funding for restoration activities. For example, organizations like the African Mangrove Initiative (AMI) work directly with local communities to restore degraded mangrove habitats, thereby contributing to both ecological resilience and the economic stability of coastal communities.

3.7 Conclusion

The study reveals several key takeaways that highlight the importance of these ecosystems for both environmental and community resilience. Firstly, mangroves play a crucial role in protecting coastal areas from erosion, storm surges, and the impacts of climate change. Their rich biodiversity supports both marine and terrestrial species, while their capacity to sequester carbon makes them vital in the fight against global warming. Additionally, mangroves provide essential resources for local communities, including fish, wood, and medicinal plants, contributing significantly to food security and livelihoods. However, the conservation of mangroves in Central Africa faces numerous challenges, including deforestation, pollution, and unsustainable exploitation. These pressures have led to the degradation of mangrove forests, impacting the ecosystem services they provide. The need for integrated, multi-stakeholder approaches to mangrove conservation is thus evident. Effective management requires collaboration among local communities, governments, NGOs, and the private sector to ensure sustainable practices are adopted and enforced. Furthermore, it is essential to incorporate traditional knowledge alongside scientific research to develop solutions that are both culturally appropriate and ecologically sound. To achieve long-term resilience, it is critical to foster community engagement and ownership of conservation efforts, empowering local populations to actively participate in restoration and sustainable management initiatives. Investments in education and capacity-building are also vital to ensure that all stakeholders are equipped with the necessary skills and knowledge. Ultimately, by working together, Central Africa's mangrove ecosystems can be safeguarded, contributing to the resilience of both the environment and the communities that depend on them for generations to come.

References

Ajonina, G. N. (2007). Mainstreaming coastal wetland biodiversity conservation in African mangroves, Cameroon. *UNDP-GEF-SFA*, 147.

Ajonina, G. N. (2008). Inventory and modelling mangrove forest stand dynamics following different levels of wood exploitation pressures in the Douala-Edea Atlantic coast of Cameroon, Central Africa. *Mitteilungen der Abteilungen für Forstliche Biometrie, Albert-Ludwigs-Universität Freiburg*, 2.

Ajonina, G. N. (2022). Cameroon Mangroves: Current status, uses, challenges, and management perspectives. In *Mangroves: biodiversity, livelihoods and conservation* (pp. 565–609). Springer Nature Singapore.

Ajonina, G. N., & Eyango, M. T. (2020). Aquaforests and aquaforestry: Africa. In *Coastal and Marine Environments* (pp. 11–40). CRC Press.

Ajonina, G. N., Aya, F. A., Diame, A., Armah, A. K., Camara, S., Amegankpoe, C., ... & Kaya, P. (2016). Overview of experience of mangrove reforestation in West and Central Africa. In *Proceedings of the 38th Annual Conference of Forestry Association of Nigeria, Port Harcourt, Rivers States, 7th-11th March* (Vol. 2016, pp. 12–21).

Ajonina, G. N., Kairo, J., Grimsditch, G., Sembres, T., Chuyong, G., & Diyouke, E. (2015). Assessment of mangrove carbon stocks in Cameroon, Gabon, the Republic of Congo (RoC) and the Democratic Republic of Congo (DRC) including their potential for reducing emissions from deforestation and forest degradation (REDD+). *The land/ocean interactions in the coastal zone of West and Central Africa*, 177–189.

Ajonina, G., Kairo, J. G., Grimsditch, G., Sembres, T., Chuyong, G., Mibog, D. E., Nyambane, A., & FitzGerald, C. (2014). *Carbon pools and multiple benefits of mangroves in Central Africa: Assessment for REDD+*, 72pp. https://www.ambienteysociedad.org.co/wp-content/uploads/2015/02/REDDcarbon_lores-1.pdf

ANPN. (2024). *Agence Nationale des Parcs Nationaux, Libreville, Gabon*. https://research-nexus.net/institution/9000044653/

Asangwe, C. K. (2006). The Douala coastal lagoon complex, Cameroon: Environmental issues. *Administering Marine Spaces: International Issues, 36*, 134–147.

Awazi, N. P., Avana-Tientcheu, M. L., Alemagi, D., Abanda, F. H., Enongene, K., Nfornkah, B. N., & Fobissie, K. (2024). Nature-based solutions for climate change adaptation and mitigation in cameroon: Realities and perspectives. *Handbook of nature-based solutions to mitigation and adaptation to climate change* (pp. 1–44). Springer International Publishing.

BBPP. (2024). *Bioko biodiversity protection program*. https://www.rainforesttrust.org/bioko-biodiversity-protection-program/

Bissonnette, J. F., Dossa, K. F., Nsangou, C. A., Satchie, Y. A., Moussa, H., Miassi, Y. E., Onguene, R., et al. (2024). What occurs within the mangrove ecosystems of the douala region in cameroon? Exploring the Challenging governance of readily available woody resources in the wouri estuary. *Environments, 11*(6), 121.

Boubakary, B., Léopold, E. K. G., Flavien, K. M. E., Maxemilie, N. M. V., Laurant, N. M., Alphonse, K. S., Din, N., et al. (2024). Growth and development of Rhizophora spp. Seedlings on Different Substrates and Insertion Level in the Wouri Estuary Mangrove (Douala, Cameroon). *Journal of Ecological Engineering, 25*(4).

Boubakary, E. K. G., Nyamsi-Moussian, L., Konango-Samè, A., Kottè-Mapoko, E. F., Motto, I. S., & Din, N. (2019). Mangrove ecosystems rehabilitation in cameroon: Effects of two abiotic factors on the growth of rhizophora seedlings. *Journal of Marine Biology and Environmental Sciences, 1*(1), 1–6.

Brunnschweiler, C., & Luisetti, T. (2022). Can blue carbon initiatives help conserve mangroves in developing countries? In *Climate and Development* (pp. 415–438).

Bumtu, K. P., Nkwatoh, A. F., & Longonje, S. N. (2020). A baseline assessment of soil organic carbon in the mangroves of the Bakassi Peninsula South-West Cameroon. *Published in Int J Trend Sci Res Dev (IJTSRD), 4*(3), 414–421.

Carrere, R. (2009). African mangroves: Their importance for people and biodiversity. *Nature & Faune, 24*(1), 3–7.

Chretien, N., Tiafack, O., & Charly, D. N. G. (2009). Mapping and monitoring urban growth on wetlands in humid tropical context using earth observation technology: case study of mangrove zones around Douala in Cameroon. In *2009 IEEE international geoscience and remote sensing symposium* (Vol. 1, pp. 1–120). IEEE.

CMN. (2024). *Cameroon mangroves and Wetlands Network: Towards a more significant position amongst actors in wetlands ecosystem conservation and more effective actions for sustainable development in municipalities.* https://www.cmrmangroves-wetlands.net/

Corcoran, E., Ravilious, C., & Skuja, M. (2007). *Mangroves of western and central Africa.* No. 26. UNEP/Earthprint.

Din, N., Ngo-Massou, V. M., Essomè-Koum, G. L., Kottè-Mapoko, E., Emane, J. M., Akongnwi, A. D., & Tchoffo, R. (2016). Local perception of climate change and adaptation in mangrove areas of the Cameroon coast. *Journal of Water Resource and Protection, 8*(5), 608–618.

Din, N., Ngo-Massou, V. M., Essomè-Koum, G. L., Ndema-Nsombo, E., Kottè-Mapoko, E., & Nyamsi-Moussian, L. (2017). Impact of urbanization on the evolution of mangrove ecosystems in the Wouri River Estuary (Douala Cameroon). *Coastal Wetlands: Alteration and Remediation,* 81–131.

Din, N., Saenger, P., Jules, P. R., Siegfried, D. D., & Basco, F. (2008). Logging activities in mangrove forests: A case study of Douala Cameroon. *African Journal of Environmental Science and Technology, 2*(2), 022–030.

Diop, E. S., Gordon, C., Semesi, A. K., Soumaré, A., Diallo, N., Guissé, A., Ayivor, J. S., et al. (2002). Mangroves of Africa. In *Mangrove ecosystems: Function and management* (pp. 63–121). Springer Berlin Heidelberg.

Diop, S., Fabres, J., Pravettoni, R., Barusseau, J. P., Descamps, C., & Ducrotoy, J. P. (2014). The western and central Africa land–sea interface: A vulnerable, threatened, and important coastal zone within a changing environment. *The land/ocean interactions in the coastal zone of west and Central Africa* (pp. 1–8). Springer International Publishing.

Ellison, J. C., & Zouh, I. (2012). Vulnerability to climate change of mangroves: Assessment from Cameroon, Central Africa. *Biology, 1*(3), 617–638.

Evariste, F. F., Jean, S. D., Victor, K., & Claudia, M. (2018). Assessing climate change vulnerability and local adaptation strategies in adjacent communities of the Kribi-Campo coastal ecosystems, South Cameroon. *Urban Climate, 24,* 1037–1051.

Feka, N. Z., & Ajonina, G. N. (2011). Drivers causing decline of mangrove in West-Central Africa: A review. *International Journal of Biodiversity Science, Ecosystem Services & Management, 7*(3), 217–230.

Feka, N. Z., & Manzano, M. G. (2008). The implications of wood exploitation for fish smoking on mangrove ecosystem conservation in the South West Province, Cameroon. *Tropical Conservation Science, 1*(3), 222–241.

Feka, Z. N., & Morrison, I. (2017). Managing mangroves for coastal ecosystems change: A decade and beyond of conservation experiences and lessons for and from west-central Africa. *Journal of Ecology and the Natural Environment, 9*(6), 99–123.

Findi, E. N., & Wantim, M. N. (2022). Using remote sensing and GIS to evaluate mangrove forest dynamics in Douala-Edea Reserve, Cameroon. *Journal of Materials and Environmental Science, 13*(3), 222–235.

Fongnzossie, E., Sonwa, D. J., Mbevo, P., Kentatchime, F., Mokam, A., Tatuebu Tagne, C., & Rim, L. F. E. A. (2022). Climate change vulnerability assessment in mangrove-dependent communities of Manoka Island, Littoral Region of Cameroon. *The Scientific World Journal, 2022*(1), 7546519.

Fonteh, M., Esteves, L. S., & Gehrels, W. R. (2009). Mapping and valuation of ecosystems and economic activities along the coast of Cameroon: Implications of future sea level rise. *Coastline Reports (EUCC International Approaches of Coastal Research in Theory and Practice), 13*, 47–63.

Fopi, R. D. T., Tchamba, M. N., & Ajonina, G. N. (2021). Caractérisation physico-chimique et dendrométrie dans les traitements de régénération de mangrove de l'Estuaire du Cameroun. *International Journal of Biological and Chemical Sciences, 15*(6), 2701–2714.

Forkam, D. C., Ajonina, G. N., Ajonina, P. U., & Tchamba, M. N. (2020). Framework for assessing the level of stakeholders involvement and governance in mangrove management: Case of selected local communities in the south west coastal Atlantic Region, Cameroon. *Journal of Ecology and the Natural Environment, 12*(4), 150–164.

Gaë, R., Musadji, N. Y., Hervé, J., Beh, M., Koutika, L. S., Ondo, J. A., & Geffroy, C. (2024). National soil organic carbon stocks inventories under different mangrove forest types in Gabon. *Open Journal of Forestry, 14*(2), 127–140.

Hou, D. (1955). Rhizophoraceae. *Flora Malesiana-Series 1, Spermatophyta, 5*(1), 429–493.

Jalloh, A., Roy-Macauley, H., & Sereme, P. (2012). Major agro-ecosystems of West and Central Africa: Brief description, species richness, management, environmental limitations and concerns. *Agriculture, Ecosystems & Environment, 157*, 5–16.

Jean-Hude, E. M., Gordon, N. A., Mbarga, A. B., & Tchikangwa, B. N. (2016). Bumpy road to improved mangrove resilience in the Douala Estuary, Cameroon. *Journal of Ecology and the Natural Environment, 8*(5), 70–89.

Kauffman, J. B., & Bhomia, R. K. (2017). Ecosystem carbon stocks of mangroves across broad environmental gradients in West-Central Africa: Global and regional comparisons. *PLoS ONE, 12*(11), Article e0187749.

Laudisoit, A., Collet, M., Muyaya, B., Mauwa, C., Ntadi, S., et al. (2017). West African Manatee Trichechus senegalensis (LINK, 1795) in the Estuary of the Congo River (Democratic Republic of the Congo): Review and Update. *J Biodivers Endanger Species, 5*, 181. https://doi.org/10.4172/2332-2543.1000181

Lonard Robert I., et al. (2020). "Biology and ecology of the halophyte Laguncularia racemosa (L.) Gaertn. f.: A review." *Handbook of Halophytes: From Molecules to Ecosystems towards Biosaline Agriculture* (2020): 1–16.

Lontsi, F. R. Z., Tchawa, P., & Mbaha, J. P. (2023). Mapping and botanical study of pressures causing mangrove dynamics of Tiko (Southwest Cameroon). *Open Access Library Journal, 10*(2), 1–19.

McGlade, J. M., Cury, P., Koranteng, K. A., & Hardman-Mountford, N. J. (2002). Environmental pollution in the Gulf of Guinea: A regional approach. *The Gulf of GLlinea Large Marine Ecosystem, 299*.

Moudingo, J. H., Ajonina, G., Kemajou, J., Wassouni, A., Bitomo, M., Assengze, A., & Tomedi, M. (2020). Sylvio-socioeconomic study of urban mangrove patches and challenges: case of Kribi, Cameroon. In *Biotechnological Utilization of Mangrove Resources* (pp. 79–115). Academic Press.

Mumbang, C., Ajonina, G. N., & Chuyong, G. B. (2025). *Four decades of cover change, degradative, and restitution stages of mangrove forest in Douala-Edea National Park, Cameroon.* https://doi.org/10.20944/preprints202501.2030.v1

Munji, C. A., Bele, M. Y., Idinoba, M. E., & Sonwa, D. J. (2014). Floods and mangrove forests, friends or foes? Perceptions of relationships and risks in Cameroon coastal mangroves. *Estuarine, Coastal and Shelf Science, 140*, 67–75.

Munji, C. A., Bele, M. Y., Nkwatoh, A. F., Idinoba, M. E., Somorin, O. A., & Sonwa, D. J. (2013). Vulnerability to coastal flooding and response strategies: The case of settlements in Cameroon mangrove forests. *Environmental Development, 5*, 54–72.

Nfotabong-Atheull, A., Din, N., & Dahdouh-Guebas, F. (2013). Qualitative and quantitative characterization of mangrove vegetation structure and dynamics in a peri-urban setting of Douala (Cameroon): An approach using air-borne imagery. *Estuaries and Coasts, 36*, 1181–1192.

Nfotabong-Atheull, A., Din, N., Essomè Koum, L. G., Satyanarayana, B., Koedam, N., & Dahdouh-Guebas, F. (2011). Assessing forest products usage and local residents' perception of environmental changes in peri-urban and rural mangroves of Cameroon, Central Africa. *Journal of Ethnobiology and Ethnomedicine, 7*, 1–13.

Ngeve, M. N., Koedam, N., & Triest, L. (2021). Genotypes of Rhizophora propagules from a non-mangrove beach provide evidence of recent long-distance dispersal. *Frontiers in Conservation Science, 2*, Article 746461.

Ngoran, S. D. (2014). *Socio-environmental impacts of sprawl on the coastline of Douala: Options for integrated coastal management*. Anchor Academic Publishing (aap_verlag).

Ngoran, S. D., Xue, X., & Ngoran, B. S. (2015). The dynamism between urbanization, coastal water resources and human health: A case study of Douala, Cameroon. *Journal of Economics and Sustainable Development, 6*(3), 167–181.

Nwabueze, B. C. (2024). *A research framework to assess the contribution of the mangrove ecosystem to the well-being of coastal communities in Africa* (Doctoral dissertation, University of British Columbia). https://open.library.ubc.ca/media/stream/pdf/24/1.0447769/4

Okafor-Yarwood, I. (2020). The effects of oil pollution on the marine environment in the Gulf of Guinea – the Bonga Oil Field example. In *Transnational Food Security* (pp. 80–97). Routledge.

Olam. (2024). *National Agency for National Parks*. https://www.olamgroup.com/content/dam/olamgroup/sustainability/sustainable-supply-chains/sustainable-palm-oil/stakeholder-engagement/stakeholder-engagement-pdfs/Response-from-the-Republic-of-Gabon-National-Agency-for-National-Parks-to-Mighty-English_RoG.pdf

Onyena, A. P., & Sam, K. (2020). A review of the threat of oil exploitation to mangrove ecosystem: Insights from Niger Delta, Nigeria. *Global Ecology and Conservation, 22*, Article e00961.

Polidoro, B. A., Ralph, G. M., Strongin, K., Harvey, M., Carpenter, K. E., Arnold, R., Williams, A., et al. (2017). The status of marine biodiversity in the Eastern Central Atlantic (West and Central Africa). *Aquatic Conservation: Marine and Freshwater Ecosystems, 27*(5), 1021–1034.

Quenum, I. A., Avocèvou-Ayisso, C., Idohou, R., Padonou, E. A., Akabassi, G. C., & Akakpo, B. A. (2024). Restoration and governance approaches of mangrove ecosystems in Africa. *Wetlands, 44*(5), 54.

Saenger, P., & Bellan, M. F. (1995). *The mangrove vegetation of the Atlantic Coast of Africa: A review*. Université de Toulouse. https://www.researchgate.net/publication/43439025_The_Mangrove_Vegetation_of_the_Atlantic_Coast_of_Africa. Accessed March 1, 2025.

Santos, M. M., Ferreira, A. V., & Lanzinha, J. C. G. (2024). *Climate change, coastal vulnerability, and mangrove protection in Africa*. Proceedings of the

International Conference on Changing Cities VI: Spatial, Design, Landscape, Heritage & Socio-economic Dimensions Rhodes Island, Greece, June 24–28, 2024.

Sereneski-Lima, C., Baggio, R. A., Pil, M. W., Boeger, M. R. T., & Boeger, W. A. (2021). Historical and contemporary factors affect the genetic diversity and structure of Laguncularia racemosa (L.) Gaertn, along the western Atlantic coast. *Estuarine, Coastal and Shelf Science, 249*, 107055.

Shokoleu, P. N., Sonwa, D. J., & Biye, E. H. (2024). Preliminary Reflection on the Opportunities and Challenges of Mangrove Ecosystem Management and Restoration in Cameroon. *ASC-TUFS Working Papers, 4*, 101–116.

4

East Africa's Coast: Bridging People and Nature for Community Resilience Through Sustainable Management of Mangrove Ecosystems

Abstract East Africa's coastal regions, encompassing Kenya, Tanzania, and Somalia, face critical socio-economic challenges such as overfishing, coastal migration, and land degradation, which threaten the resilience of local communities. Mangrove ecosystems play a vital role in addressing these challenges, providing critical services such as coastal protection, fisheries support, and biodiversity conservation. However, the rapid deforestation of mangroves due to human activities, exacerbated by climate change, has led to significant socio-economic impacts, affecting fisheries, tourism, and the natural defense against coastal erosion. This study explores the ecological and socio-economic importance of mangroves, focusing on their role in enhancing community resilience. It highlights successful community-based conservation initiatives in the region, emphasizing the importance of collaboration between local communities, governments, and NGOs. By integrating traditional knowledge with scientific research, the study proposes a framework for sustainable mangrove management. Key policy recommendations are presented, advocating for the inclusion of mangrove protection within broader regional and national development strategies. The study underscores the necessity of bridging people and nature for the long-term sustainability of East Africa's coastal ecosystems, offering valuable insights into fostering resilience through effective mangrove conservation.

Keywords East Africa · Mangrove ecosystems · Community resilience · Coastal challenges · Sustainable management · Climate change

4.1 Overview of East Africa's Coastal Challenges

East Africa is home to some of the most diverse and dynamic coastal regions in the world, with countries like Kenya, Tanzania, Seychelles, and Somalia possessing a significant stretch of coastline along the Indian Ocean (Mudoola, 2021; Samoilys et al., 2015). The coastal zone of this region is vital for the livelihoods of millions of people, contributing to fisheries, agriculture, and tourism. However, it also faces numerous environmental and socio-economic challenges that are exacerbated by both natural and human-induced factors (Diop & Scheren, 2016; Pollard, 2012). Kenya, Tanzania, and Somalia are among the most populous countries in East Africa, and their coastal regions are pivotal to their economies. Kenya's coastline, stretching for about 536 kilometers, is renowned for its vibrant fishing industry, beautiful beaches, and rich biodiversity, particularly in marine ecosystems like coral reefs and mangroves (Busiega & Busiega, 2016; Paula, 2016). Tanzania's coastline is even more expansive, with 800 kilometers of shoreline, and includes the Zanzibar Archipelago, which plays a critical role in tourism and marine resources (Gates et al., 2021; Gustavson et al., 2009; Lödel, 2025; Muhando & Rumisha, 2008). Somalia, despite the challenges of political instability, also has a long coastline that is key for both trade and fishing (Farah, 2016). Coastal regions in these countries are home to many communities whose livelihoods depend on the natural resources of the ocean, such as artisanal fishing, agriculture, and tourism (Holly et al., 2022).

However, these coastal areas are experiencing significant socio-economic and environmental challenges. One of the most pressing issues is overfishing. The rapid increase in demand for fish and seafood, coupled with unsustainable fishing practices, has led to the depletion of marine resources in the region. Overfishing not only threatens biodiversity but also undermines the livelihoods of coastal communities that depend on fishing as their primary source of income (Doerr, 2016). In Kenya, for instance, the government has implemented policies to address overfishing, but the issue remains widespread, exacerbated by illegal, unreported, and unregulated (IUU) fishing (Kamau et al., 2009).

Coastal migration is another significant socio-economic challenge. Climate change, coupled with political instability, has caused many people to migrate from inland regions to coastal cities in search of better opportunities. This has led to overcrowded urban areas, putting a strain on public services and infrastructure. In Somalia, for example, large numbers of people from drought-prone inland areas have moved to coastal cities like Mogadishu, exacerbating the pressures on the local economy and the environment (Abdullahi et al., 2025). As populations increase, the need for more land and resources leads to land degradation, as coastal areas are often cleared for settlement or agricultural purposes, destroying natural habitats and further threatening marine ecosystems.

In addressing these challenges, mangroves have emerged as crucial ecosystems for mitigating environmental degradation. Mangrove forests are vital in stabilizing shorelines, preventing erosion, and acting as natural barriers against coastal flooding and storm surges, which are becoming more frequent due to climate change. Studies have shown that mangroves play a key role in carbon sequestration, making them valuable not only for environmental protection but also in the fight against global warming (Hamza et al., 2020). In Kenya, community-based initiatives to restore mangrove ecosystems have proven effective in boosting fish populations and improving the resilience of coastal communities against the effects of climate change (Kiprono, 2021). These efforts are helping to address overfishing and coastal land degradation, while also providing alternative livelihoods through sustainable ecotourism and sustainable fish harvesting.

4.2 Socio-Economic Impacts of Mangrove Loss in East Africa

The loss of mangrove forests in East Africa has significant socio-economic repercussions for local communities and economies. Mangroves, which are found along the coastal regions of countries such as Kenya, Tanzania, and Mozambique, play a critical role in sustaining biodiversity, supporting local livelihoods, and protecting coastal ecosystems (Table 4.1). However, widespread mangrove deforestation is increasingly threatening these benefits, leading to devastating consequences for the socio-economic fabric of these coastal communities.

In East Africa, mangroves provide essential services to coastal communities, particularly those dependent on agriculture, fisheries, and tourism. Local populations often rely on mangrove ecosystems for their livelihoods, including the collection of wood for fuel, harvesting of honey, and gathering of fish and crabs. The loss of mangrove habitats directly undermines these sources of income. According to Brander et al. (2012) and Huxham et al. (2017), mangroves are estimated to generate billions of dollars in ecosystem services globally, but in East Africa, this contribution is often overlooked. When mangroves are cleared for development or agriculture, these communities face diminished resources, leading to economic hardship and poverty. In particular, women, who are often responsible for collecting fuelwood and small marine products, are disproportionately affected by mangrove degradation (Nyangoko et al., 2022).

Mangrove ecosystems distributed widely across the East African coasts, play a pivotal role in supporting fisheries by serving as nurseries for various fish species, which depend on the sheltered waters of mangrove forests during their juvenile stages. The loss of these critical habitats leads to declines in fish populations, which, in turn, negatively impacts local fisheries. The East African region is home to many small-scale fishers who rely on the availability of fish for both sustenance and income. As fish stocks decrease due to mangrove loss, these fishers face reduced catches, leading to economic losses and food insecurity (Makame et al., 2015; Thoya et al., 2022). Tourism is another sector severely impacted by mangrove loss. Mangrove forests, with their rich biodiversity, attract

Table 4.1 Mangrove species in East Africa

Mangrove species	Description	References
Avicennia marina (Grey mangrove)	*Avicennia marina*, also known as the grey mangrove, is one of the most prevalent mangrove species in East Africa. This species is distinguished by salt-excreting glands on its leaves and is highly adapted to saline environments. The grey mangrove often forms dense clusters in the intertidal zone, playing a vital role in stabilizing sediments and preventing coastal erosion. Its buoyant propagules, which can remain viable for months, aid in its extensive spread throughout the estuaries and coastal regions of the area	Triest et al. (2020), Njana (2020), Triest et al. (2021a), and Amade et al. (2021)
Rhizophora mucronata (Red mangrove)	*Rhizophora mucronata*, commonly referred to as the red mangrove, is recognized for its unique prop roots, which provide support in soft, muddy soils. These roots also form intricate habitats for numerous marine species. Red mangroves are especially abundant in the tidal zones and estuaries of East Africa, where their dense root networks help shield the coastline from erosion and storm surges. The propagules of red mangroves are viviparous, meaning they start to germinate while still attached to the parent tree, facilitating their successful establishment in new locations	Edith et al. (2020), Oloo (2020), Triest et al. (2021b), Waweru et al. (2022), and Njiru et al. (2022)

(continued)

Table 4.1 (continued)

Mangrove species	Description	References
Bruguiera gymnorrhiza (Large-leafed mangrove)	*Bruguiera gymnorrhiza*, also called the large-leafed mangrove, is distinguished by its wide, leathery leaves and unique knee-like aerial roots. This species is typically found in the upper intertidal zones of East African mangrove ecosystems, where it helps stabilize the soil and prevent erosion. The large-leafed mangrove also provides habitat for various animals, such as crabs and mollusks, which play a role in its ecological interactions	Nabeelah Bibi et al. (2019), Musara et al. (2020), and Li et al. (2021)
Ceriops tagal (Stilted mangrove)	*Ceriops tagal*, known as the stilted mangrove, is recognized for its stilt-like roots that offer stability in the shifting sediments of intertidal areas. This species thrives in slightly lower salinity conditions compared to some other mangrove species and is commonly found in mixed mangrove forests. Ceriops tagal plays an important role in the mangrove ecosystem by promoting sediment accumulation and offering habitat for various marine species	Smith (2019), Mburu (2019), Raju (2019), and Manohar et al. (2023)
Sonneratia alba (White mangrove)	*Sonneratia alba*, also known as the white mangrove, is recognized for its large, fleshy fruit and distinct flower structure. This species is usually found in the upper intertidal zones and is distinguished by its buttressed roots, which help stabilize the tree and trap sediment. The white mangrove plays a key role in supporting biodiversity, providing shelter for birds, insects, and other wildlife, and contributing to the overall health of mangrove ecosystems	Jenoh et al. (2019), Edith et al. (2020), Warui et al. (2020), Jenoh et al. (2021), Kiti et al. (2021), and Okello et al. (2024)

tourists interested in birdwatching, fishing, and ecotourism. The degradation of these coastal ecosystems can lead to a decline in tourist numbers, thus undermining local economies that depend on this sector. Moreover, mangroves act as a natural barrier against coastal erosion and storm surges, protecting infrastructure, agricultural lands, and settlements. The absence of mangroves leads to increased vulnerability to natural disasters such as floods and cyclones, which can cause significant economic damage to coastal communities.

The effects of climate change, particularly rising sea levels and more frequent and intense storms, further exacerbate the vulnerabilities of coastal communities in East Africa. Mangroves, with their ability to sequester carbon and stabilize coastlines, play an important role in mitigating climate change impacts. However, the destruction of these forests reduces the region's resilience to the consequences of climate change. Coastal erosion becomes more pronounced, and the risk of flooding increases, which threatens both human life and infrastructure. Furthermore, as mangroves disappear, their role in carbon sequestration is lost, contributing to the acceleration of climate change (Leal Filho et al., 2021).

In East Africa, several mangrove species play crucial roles in coastal ecosystems (Table 4.1). *Avicennia marina* (Grey Mangrove) is widespread and adapted to high salinity environments. Its salt-excreting glands and buoyant propagules allow it to thrive in intertidal zones, stabilizing sediments and reducing erosion (Amade et al., 2021; Njana, 2020; Triest et al., 2020, 2021a). *Rhizophora mucronata* (Red Mangrove) features distinctive prop roots that provide stability in muddy substrates. It helps prevent coastal erosion and supports marine species, with viviparous propagules aiding its spread (Edith et al., 2020; Njiru et al., 2022; Oloo, 2020; Triest et al., 2021b; Waweru et al., 2022). *Bruguiera gymnorrhiza* (Large-Leafed Mangrove) is recognized for its broad leaves and knee-like aerial roots, stabilizing soils and supporting diverse fauna in higher intertidal zones (Li et al., 2021; Musara et al., 2020; Nabeelah Bibi et al., 2019). *Ceriops tagal* (Stilted Mangrove) has stilt-like roots that provide stability in shifting sediments, thriving in less saline conditions. It contributes to sediment accretion and offers habitat for marine organisms (Manohar et al., 2023; Mburu, 2019; Raju, 2019; Smith,

2019). *Sonneratia alba* (White Mangrove) is known for its large fruit and unique flowers, occupying higher intertidal zones with buttressed roots for stability and sediment trapping. It supports biodiversity, providing habitats for various species (Edith et al., 2020; Jenoh et al., 2019, 2021; Kiti et al., 2021; Okello et al., 2024; Warui et al., 2020). These species collectively enhance coastal resilience and biodiversity in East Africa.

4.3 Community-Based Mangrove Conservation Initiatives in East Africa

Mangrove ecosystems play a crucial role in protecting coastal areas from erosion, supporting biodiversity, and acting as carbon sinks. In East Africa, mangrove conservation has gained significant attention due to their rapid depletion caused by human activities, including overharvesting, land reclamation, and pollution. However, various successful community-based initiatives across Kenya, Tanzania, Somalia, and other East African nations have emerged, emphasizing the importance of local involvement in sustainable resource management. These initiatives rely on collaborative approaches between local communities, governments, and non-governmental organizations (NGOs), empowering people to manage their mangrove resources for long-term sustainability.

In Kenya, one of the most notable examples of community-based mangrove conservation is the Kisite-Mpunguti Marine National Park and Reserve. This project, driven by the Kenya Wildlife Service (KWS) and local communities, involves the restoration of degraded mangrove areas. The Pate Island Mangrove Restoration Project, as well as the Lamu and Tana mangrove restoration projects led by the Coast Development Authority (CDA) and local community groups, has focused on replanting mangroves, improving livelihoods, and ensuring sustainable resource use. According to PANORAMA (2025) and CDA (2025), the restoration efforts have shown promising results, with significant increases in mangrove density and species diversity. Additionally, the involvement of local communities in monitoring and managing mangrove areas has reduced illegal harvesting, ensuring a more sustainable future for the ecosystem. Similarly, in Tanzania, the Mangrove

Forest Reserve project, spearheaded by the Tanzania Forest Service (TFS) Agency and supported by international organizations such as the World Bank, has been instrumental in enhancing mangrove conservation through community engagement (TFS, 2025). The Zanzibar Mangrove Conservation and Restoration Project under the coordination of the Zanzibar Association for Climate Change Resilience (ZACCR) has effectively combined community-based approaches with governmental policies, where local villagers, particularly women, are involved in monitoring, restoration, and sustainable utilization of mangrove resources (International Tree Foundation, 2025). This project supported by the International Tree Foundation, UK, emphasizes the sustainable harvesting of mangrove products like wood and fish, offering an income alternative to destructive practices. In Somalia, where political instability has posed significant challenges, local communities, NGOs, and international organizations have collaborated to protect mangrove forests, especially in South Central Somalia, Puntland, and Somaliland (ASCLME, 2012; Mumuli et al., 2010; UNCCD, 2016). Mangroves conservation and restoration initiatives in Somalia such as the Green Somalia Initiative (GSI) have facilitated community-led initiatives, where communities are empowered to restore mangrove habitats through capacity-building, training, and developing sustainable livelihoods like ecotourism and fish farming (GSI, 2025). According to Little (2018), these efforts have led to improved mangrove cover and enhanced resilience against coastal erosion, highlighting the importance of community engagement in fragile and conflict-prone regions.

The success of these projects is largely due to the collaborative approaches between local communities, governments, and NGOs. In many cases, local communities possess intimate knowledge of the ecosystems, and their participation is critical for long-term sustainability. Governments and NGOs bring technical expertise, resources, and policy frameworks to ensure effective management and protection of mangrove forests. For example, Tanzania's National Forest Policy encourages community involvement in forest management and provides a legal framework for community forestry initiatives. Empowering local communities has been central to the success of these conservation projects (National Forest Policy, 1998). By giving communities ownership of mangrove resources, they are more likely to adopt sustainable practices. Furthermore, income-generating activities, such as ecotourism, sustainable fisheries, and carbon trading, offer incentives for conservation. As noted by Henderson et al. (2021), the integration of local knowledge with scientific methods has led to more resilient and adaptive conservation strategies, ensuring that mangrove ecosystems are protected for future generations.

4.4 Bridging People and Nature: Integrating Local Knowledge with Science for Sustainable Mangrove Management in East Africa

Mangrove ecosystems in East Africa provide critical services, including coastal protection, biodiversity conservation, and livelihoods for local communities. These unique ecosystems, however, face increasing threats from climate change, deforestation, and urbanization. Effective management of mangroves requires a holistic approach that integrates both scientific research and traditional knowledge to ensure sustainability. Bridging the gap between these two knowledge systems is crucial for preserving these valuable ecosystems and ensuring their long-term survival.

Traditional knowledge, passed down through generations, is deeply embedded in the cultural practices and worldviews of local communities. This knowledge often includes sustainable management practices, such as controlled harvesting and seasonal migration, which have helped preserve mangrove ecosystems for centuries. For instance, coastal communities in Kenya and Tanzania have long understood the importance of mangroves for protecting shorelines and supporting fisheries (Ahmed et al., 2022; Mangora et al., 2016). Scientific research, on the other hand, provides detailed data on mangrove ecology, carbon sequestration potential, and the impacts of climate change. Integrating these two knowledge systems enhances the understanding of mangrove ecosystems. Combining scientific research with local knowledge can improve the effectiveness of conservation efforts and ensure that management strategies are contextually appropriate. According to Mose et al. (2018), the integration of local and scientific knowledge creates a more comprehensive framework for resource management, where the community's experiences and the scientific community's findings complement each other. This synergy fosters mutual respect and increases the likelihood of success in managing mangroves sustainably.

Community involvement is fundamental to the long-term sustainability of mangrove management. To encourage local participation, it is essential to recognize and validate local knowledge systems while incorporating them into decision-making processes. In East Africa, community-based management approaches are being increasingly implemented to involve local stakeholders in mangrove conservation. For example, participatory mapping and community-driven monitoring programs have been successful in Kenya, where local communities have worked alongside scientists to monitor mangrove health and report changes (Huff & Tonui, 2017; Vallerani, 2017). These initiatives empower local communities to actively manage and protect mangroves while respecting traditional knowledge. Furthermore, fostering strong local leadership and ensuring that mangrove management decisions are locally driven can improve the outcomes of conservation efforts. A key aspect of this approach is the creation of local management committees or community conservancies, which provide a platform for dialogue between scientists, community members, and other stakeholders. Engaging women and youth in these initiatives is also crucial for ensuring that the benefits of mangrove management are equitably distributed and that future generations are prepared to continue conservation efforts.

Capacity-building is critical for ensuring that local stakeholders have the skills and resources to effectively manage mangrove ecosystems. In East Africa, several programs have been established to enhance the capacity of local communities, government authorities, and nongovernmental organizations. For example, Mangrove Restoration Initiatives in Zanzibar championed by the Zanzibar Association for Climate

Change Resilience (ZACCR) have focused on training local communities in sustainable mangrove restoration techniques and providing them with tools to monitor mangrove health (Mohamed et al., 2023; Staehr et al., 2018). These capacity-building initiatives promote knowledge exchange between local people and researchers, fostering a collaborative environment for mangrove management. Investing in education and training at all levels ensures that local stakeholders are equipped to address challenges such as climate change and overexploitation. Additionally, local community leaders can act as change agents, transferring skills and knowledge to others, thereby expanding the impact of conservation efforts. As noted by Jape and Najar (2024), capacity-building efforts should prioritize empowering local communities with the knowledge to sustainably manage their natural resources, ensuring that they become active participants in the decision-making processes.

4.5 Policy Recommendations for Sustainable Mangrove Management in East Africa

Mangroves are vital ecosystems providing numerous environmental, economic, and social benefits, including coastal protection, carbon sequestration, and support for biodiversity. In East Africa, countries such as Kenya, Tanzania, Seychelles, and Somalia are home to rich mangrove forests that support local communities and ecosystems. However, these mangroves are under threat from unsustainable practices like logging, land conversion for agriculture, and climate change impacts. Effective policy frameworks at national, local, and regional levels are crucial for their sustainable management.

National policies are essential in ensuring the protection and restoration of mangrove ecosystems. In East Africa, there is a need for comprehensive policies that recognize the importance of mangroves in national development plans. For example, Tanzania has a Mangrove Management Plan that emphasizes the sustainable use and conservation of mangrove forests (Mangrove Alliance, 2022). However, gaps remain in implementation, enforcement, and integrating mangrove conservation into wider development agendas. Local policies, including community-based management approaches, are critical to ensure that mangrove restoration aligns with the needs and interests of local populations. In Kenya, the involvement of local communities in mangrove restoration projects, such as the Mangrove Restoration Initiative in Lamu, has demonstrated the success of community-driven management. These local policies should foster partnerships between communities, local governments, and conservation organizations. The inclusion of indigenous knowledge and the recognition of community tenure rights can enhance the success of mangrove restoration efforts (Kairu et al., 2024).

Regional cooperation among East African countries is vital for addressing the transboundary challenges facing mangrove ecosystems. Mangroves do not respect national boundaries, and issues such as climate change and sea-level rise require coordinated efforts across borders. One example of regional cooperation is the East African Marine and Coastal Environment Management Project (MACEMP), which supports integrated coastal zone management, including mangrove protection, across Kenya, Tanzania, and Zanzibar. International frameworks such as the Ramsar Convention on Wetlands and the United Nations Environment Programme's (UNEP) Regional Seas Programme provide a platform for collaboration. These frameworks encourage the signatory countries to adopt national policies for the protection of wetland ecosystems, including mangroves, and provide funding for restoration projects. The adoption of the Paris Agreement on climate change also highlights the importance of coastal ecosystems like mangroves in mitigating climate impacts. East African nations should work toward enhancing their commitments to international frameworks, strengthening cross-border collaboration for mangrove restoration and conservation.

Integrating mangrove protection into broader development strategies is essential for ensuring long-term sustainability. One key lesson is the need for policies that address the balance between development and conservation. In many East African nations, economic pressures from agriculture, tourism, and infrastructure development pose significant threats to mangrove habitats. Policies should incentivize sustainable land-use practices that reduce the encroachment of development activities into mangrove areas. A successful approach can be seen in the case of Zanzibar, Tanzania, Kenya, and Seychelles, where the government has integrated mangrove conservation into national climate change adaptation strategies. This approach not only protects the environment but also supports livelihoods, particularly for communities reliant on fisheries and ecotourism. Similarly, financial mechanisms such as Payments for Ecosystem Services (PES) can provide incentives for local communities to conserve mangroves while benefiting from sustainable practices (Osewe et al., 2024). Moreover, mangrove protection must be viewed as part of the broader framework for achieving the United Nations Sustainable Development Goals (SDGs), particularly those related to life below water, climate action, and sustainable livelihoods.

4.6 Conclusion

Bridging people and nature is fundamental to the sustainable management of mangrove ecosystems in East Africa. Mangroves serve as crucial buffers against coastal erosion, provide vital ecosystem services, and are home to diverse marine life, yet they remain under threat from human activities such as deforestation, overexploitation, and climate change. To ensure their preservation and strengthen community resilience, a holistic approach that integrates local communities in conservation efforts is essential. The involvement of local populations not only fosters a deeper connection between people and their natural environment but also empowers communities to take ownership of their resources, leading to more effective and lasting conservation outcomes. Strategically, enhancing community resilience in East Africa requires a multifaceted approach. First, strengthening community-based conservation initiatives is essential. This can be achieved by increasing awareness about the vital roles of mangroves, particularly through educational campaigns that highlight their ecological and economic benefits. Second, capacity-building programs should be implemented to equip communities with the necessary skills to manage and protect mangrove resources effectively. This includes training in sustainable harvesting practices, ecotourism development, and mangrove restoration techniques. Moreover, fostering partnerships between local communities, governments, non-governmental organizations, and the private sector can help create a collaborative framework for sustainable mangrove management. These partnerships can secure funding, policy support, and technical expertise needed to scale conservation efforts. Lastly, integrating mangrove conservation into national and regional climate change adaptation strategies will ensure that mangrove ecosystems are prioritized in future development plans. Ultimately, bridging people and nature through education, collaboration, and capacity-building is critical to enhancing community resilience and safeguarding East Africa's mangrove ecosystems for generations to come.

References

Abdullahi, S., Singh, R., Takaindisa, J., Giacomelli, C., Sax, N., Carneiro, B., & Pacillo, G. (2025). *The nexus between climate change, mobility, and conflict in Somalia* (Discussion Paper 2025/03). 35 p. https://hdl.handle.net/10568/173200

Ahmed, H. A., Mwaura, F., Thenya, T., & Kairo, J. G. (2022). Coastal and mangrove economic valuation associated fisheries and problems in Kwale County, Kenya. *Indo Pacific Journal of Ocean Life, 6*(1).

Amade, F. M., Oosthuizen, C. J., & Chirwa, P. W. (2021). Genetic diversity and contemporary population genetic structure of Avicennia marina from Mozambique. *Aquatic Botany, 171*, Article 103374.

ASCLME (2012). *National marine ecosystem diagnostic analysis. Somalia. Contribution to the Agulhas and Somali current large marine ecosystems project* (supported by UNDP with GEF grant financing). https://nairobiconvention.org/clearinghouse/sites/default/files/National%20Marine%20Ecosystem%20Diagnostic%20Analysis%20%28MEDA%29%20-%20Somalia.pdf

Brander, L. M., Wagtendonk, A. J., Hussain, S. S., McVittie, A., Verburg, P. H., de Groot, R. S., & van der Ploeg, S. (2012). Ecosystem service values for mangroves in Southeast Asia: A meta-analysis and value transfer application. *Ecosystem Services, 1*(1), 62–69.

Busiega, J. N., & Busiega, J. N. (2016). *Harnessing maritime security and resource exploitation: Role of maritime diplomacy in Kenya* (Doctoral dissertation, University Of Nairobi).

CDA. (2025). *Coast development authority.* https://cda.go.ke/?p=4109

Diop, S., & Scheren, P. A. (2016). Sustainable oceans and coasts: Lessons learnt from Eastern and Western Africa. *Estuarine, Coastal and Shelf Science, 183*, 327–339.

Doerr, F. (2016). *Blue growth and ocean grabbing: A historical materialist perspective on fisheries in East Africa. In An international colloquium Global governance/politics, climate justice & agrarian/social justice: linkages and challenges.* Colloquium Paper (Vol. 18).

Edith, M. M., Huxley, M. M., Chinedu, C. O., Joyce, M. J., James, H. P., & Damase, P. K. (2020). Isolation, characterization and biotechnological potential of tropical culturable rhizospheric fungi from four mangrove species in Kenya. *African Journal of Microbiology Research, 14*(9), 541–554.

Farah, Q. H. (2016). *The stability/sustainability dynamics: The case of marine environmental management in Somalia.* https://yorkspace.library.yorku.ca/bitstreams/e970f139-0439-4acf-97c8-55f9e4e744b6/download

Gates, A. R., Durden, J. M., Richmond, M. D., Muhando, C. A., Khamis, Z. A., & Jones, D. O. B. (2021). Ecological considerations for marine spatial management in deep-water Tanzania. *Ocean & Coastal Management, 210*, Article 105703.

GSI. (2025). *Green Somalia initiative.* https://www.greensomalinitiative.org/

Gustavson, K., Kroeker, Z., Walmsley, J., & Juma, S. (2009). A process framework for coastal zone management in Tanzania. *Ocean & Coastal Management, 52*(2), 78–88.

Hamza, A. J., Esteves, L. S., Cvitanovic, M., & Kairo, J. (2020). Past and present utilization of mangrove resources in Eastern Africa and drivers of change. *Journal of Coastal Research, 95*(SI), 39–44.

Henderson, J., Breen, C., Esteves, L., La Chimia, A., Lane, P., Macamo, S., Wynne-Jones, S., et al. (2021). Rising from the depths Network: A challenge-led research agenda for marine heritage and sustainable development in Eastern Africa. *Heritage, 4*(3), 1026–1048.

Holly, G., Rey da Silva, A., Henderson, J., Bita, C., Forsythe, W., Ombe, Z. A., Roberts, H., et al. (2022). Utilizing marine cultural heritage for the preservation of coastal systems in East Africa. *Journal of Marine Science and Engineering, 10*(5), 693.

Huff, A., & Tonui, C. (2017). *Making 'mangroves together': Carbon, conservation and co-management in Gazi Bay, Kenya*. https://opendocs.ids.ac.uk/ndownloader/files/48228814

Huxham, M., Dencer-Brown, A., Diele, K., Kathiresan, K., Nagelkerken, I., & Wanjiru, C. (2017). Mangroves and people: Local ecosystem services in a changing climate. *Mangrove Ecosystems: A Global Biogeographic Perspective: Structure, Function, and Services, 245–274*.

International Tree Foundation. (2025). *Building community resilience through mangrove restoration in Zanzibar*. https://www.internationaltreefoundation.org/news/building-community-resilience-through-mangrove-restoration-in-zanzibar

Jape, K. K., & Najar, M. A. (2024). Empowering local stewardship of coastal ecosystems in Zanzibar: Participatory models for habitat rehabilitation and resilience. *International Journal of Advanced Multidisciplinary Research, 4*(2), 185–194.

Jenoh, E. M., de Villiers, E. P., de Villiers, S. M., Okoth, S., Jefwa, J., Kioko, E., ... & Koedam, N. (2019). Infestation mechanisms of two woodborer species in the mangrove Sonneratia alba J. Smith in Kenya and co-occurring endophytic fungi. *Plos one, 14*(10), e0221285.

Jenoh, E. M., Traoré, M., Kosore, C., & Koedam, N. (2021). Biochemical response of Sonneratia alba Sm. branches infested by a wood boring moth (Gazi Bay, Kenya). *PLoS One, 16*(11), e0259261.

Kairu, A., Mbeche, R., Kotut, K., & Kairo, J. (2024). From centralization to decentralization: Evolution of forest policies and their implications on mangrove management in Kenya. *Forest Policy and Economics, 168*, Article 103290.

Kamau, E. C., Wamukota, A., & Muthiga, N. (2009). Promotion and management of marine fisheries in Kenya. *Towards Sustainable Fisheries Law*, 83.

Kiprono, A. (2021). *An assessment of the effectiveness of mangrove restoration projects along the Kenyan coast* (Doctoral dissertation, University of Nairobi). https://erepository.uonbi.ac.ke/bitstream/handle/11295/155712/Kiprono%20_An%20Assessment%20of%20the%20Effectiveness%20of%20Mangrove%20Restoration%20Projects%20Along%20the%20Kenyan%20Coast.pdf?sequence=1&isAllowed=y

Kiti, H. M., Munga, C. N., Odalo, J. O., Guyo, P. M., & Kibiti, C. M. (2021). Diversity of mangrove fungal endophytes from selected mangrove species of coastal Kenya. *Western Indian Ocean Journal of Marine Science, 20*(1), 125–136.

Leal Filho, W., Azeiteiro, U. M., Balogun, A. L., Setti, A. F. F., Mucova, S. A., Ayal, D., Oguge, N. O., et al. (2021). The influence of ecosystems services depletion to climate change adaptation efforts in Africa. *Science of the Total Environment, 779*, Article 146414.

Li, S. Y., Li, Y. C., Zhang, T. H., Qin, L. L., An, Y. D., Pang, Y. K., & Jiang, G. F. (2021). Characterization of the complete chloroplast genome of mangrove Bruguiera gymnorrhiza (L.) Lam. ex Savigny. *Mitochondrial DNA Part B, 6*(7), 2076–2078.

Little, D. I. (2018). Mangrove restoration and mitigation after oil spills and development projects in East Africa and the Middle East. *Threats to Mangrove Forests: Hazards, Vulnerability, and Management*, 637–698.

Lödel, M. (2025). *Interactions between tourism and small-scale fisheries in Zanzibar: An integrated socio-ecological study.*

Makame, M. O., Kangalawe, R. Y., & Salum, L. A. (2015). Climate change and household food insecurity among fishing communities in the eastern coast of Zanzibar. *Journal of Development and Agricultural Economics, 7*(4), 131–142.

Mangora, M. M., Lugendo, B. R., Shalli, M. S., & Semesi, S. (2016). Mangroves of Tanzania. *Mangroves of the Western Indian Ocean: status and management*, 33–49.

Mangrove Alliance (2022). *How can policy and legal frameworks be strengthened to ensure sustainable use and conservation of mangroves?* Tanzania Policy Brief. https://www.mangrovealliance.org/wp-content/uploads/2020/03/PB_tanzania_updatedOct2022-long-regional-text.pdf

Manohar, S. M., Yadav, U. M., Kulkarnii, C. P., & Patil, R. C. (2023). *An overview of the phytochemical and pharmacological profile of the spurred mangrove Ceriops tagal (Perr.)* CB Rob.

Mburu, F. (2019). *Tridecadal assessment of mangrove cover and cover change in the trans boundary areas of Kenya and Tanzania between 1986–2018* (Doctoral dissertation, Egerton University).

Mohamed, M. K., Adam, E., & Jackson, C. M. (2023). Policy review and regulatory challenges and strategies for the sustainable mangrove management in Zanzibar. *Sustainability, 15*(2), 1557.

Mose, V. N., Western, D., & Tyrrell, P. (2018). Application of open source tools for biodiversity conservation and natural resource management in East Africa. *Ecological Informatics, 47*, 35–44.

Mudoola, D. (2021). *Blue economy and integrated transit route in East African Community* (Master's thesis, Ankara Universitesi (Turkey)). https://www.proquest.com/openview/88bf4dc8322fa5debd0f27fa4211570b/1?pq-origsite=gscholar&cbl=2026366&diss=y

Muhando, C. A., & Rumisha, C. K. (2008). Distribution and status of coastal habitats and resources in Tanzania. *Institute for Marine Sciences, University of Dar es Salaam.* https://citeseerx.ist.psu.edu/document?repid=rep1&type=pdf&doi=7a2a0900b5930df2385454b6eefbfc6b6127192f

Mumuli, S. O, Alim, M., & Oduori, G. (2010). *Monitoring of mangroves in Somalia* (Puntland, Somaliland and South-Central Somalia). FAO-SWALIM. Project Report No. L-19. Nairobi, Kenya. https://www.nairobiconvention.org/clearinghouse/sites/default/files/Monitoring%20of%20Mangroves%20in%20Somalia%20%28Puntland%2C%20Somaliland%20and%20South%20Central%20Somalia%29.pdf

Musara, C., Aladejana, E. B., & Mudyiwa, S. M. (2020). Review of botany, nutritional, medicinal, pharmacological properties and phytochemical constituents of bruguiera gymnorhiza (L.) Lam (Rhizophoraceae). *Journal of Pharmacy and Nutrition Sciences, 10*(4), 123–132.

Nabeelah Bibi, S., Fawzi, M. M., Gokhan, Z., Rajesh, J., Nadeem, N., RR, R. K., Pandian, S. K., et al. (2019). Ethnopharmacology, phytochemistry, and global distribution of mangroves—A comprehensive review. *Marine Drugs, 17*(4), 231.

National Forest Policy. (1998). https://faolex.fao.org/docs/pdf/tan175173.pdf

Njana, M. A. (2020). Structure, growth, and sustainability of mangrove forests of mainland Tanzania. *Global Ecology and Conservation, 24*, Article e01394.

Njiru, D. M., Githaiga, M. N., Nyaga, J. M., Lang'at, K. S., & Kairo, J. G. (2022). Geomorphic and climatic drivers are key determinants of structural variability of mangrove forests along the Kenyan Coast. *Forests, 13*(6), 870.

Nyangoko, B. P., Berg, H., Mangora, M. M., Shalli, M. S., & Gullström, M. (2022). Local perceptions of changes in mangrove ecosystem services and their implications for livelihoods and management in the Rufiji Delta, Tanzania. *Ocean & Coastal Management, 219*, Article 106065.

Okello, J. A., Koedam, N., Di Nitto, D., Dahdouh-Guebas, F., Van der Stocken, T., Hugé, J., Suarez, E., et al. (2024). *IUCN red list of ecosystems, mangroves of the Western Indian Ocean.*

Oloo, C. (2020). *Comparative sorption of Organic Dyes using Xylocarpus Moluccensis and Rhizophora Mucronata Mangrove species from Kenyan Coastal Region* (Doctoral dissertation, University of Nairobi).

Osewe, E. O., Popa, B., Vacik, H., Osewe, I., & Abrudan, I. V. (2024). Review of forest ecosystem services evaluation studies in East Africa. *Frontiers in Ecology and Evolution, 12*, 1385351.

PANORAMA. (2025). *Roots of change: Community-based ecological mangrove restoration in Kenya*. https://panorama.solutions/en/solution/roots-change-community-based-ecological-mangrove-restoration-kenya

Paula, J. (2016). Overall assessment of the state of the coast in the Western Indian Ocean. *Western Indian Ocean*, 501.

Pollard, E. (2012). Present and past threats and response on the east coast of Africa: An archaeological perspective. *Journal of Coastal Conservation, 16*, 143–158.

Raju, A. S. (2019). *Status of pollination ecology studies on mangroves. Advances in Pollen Spore Research*. Today & Tomorrow's Printers and Publishers.

Samoilys, M., Pabari, M., Andrew, T., Maina, G. W., Church, J., Momanyi, A., . Mutta, D., et al. (2015). Resilience of coastal systems and their human partners in the Western Indian Ocean. In *Nairobi: IUCN ESARO, WIOMSA, CORDIO, UNEP Nairobi Convention*.

Smith, S. A. (2019). *Vulnerability and adaptability of mangrove forests on Misali Island, Zanzibar, Tanzania*.

Staehr, P. A., Sheikh, M., Rashid, R., Ussi, A., Suleiman, M., Kloiber, U., Muhando, C., et al. (2018). Managing human pressures to restore ecosystem health of Zanzibar coastal waters. *J. Aquac. Mar. Biol, 7*(2), 59–70.

TFS. (2025). *Tanzania Forest Service—Restoration of mangrove ecosystem in Tanzania for enhancement of local communities*. https://sdgs.un.org/partnerships/restoration-mangrove-ecosystem-tanzania-enhancement-local-communities#:~:text=The%20objective%20of%20this%20commitment,transfer%20of%20marine%20technology%20by

Thoya, P., Horigue, V., Möllmann, C., Maina, J., & Schiele, K. S. (2022). Policy gaps in the East African Blue economy: Perspectives of small-scale fishers on port development in Kenya and Tanzania. *Frontiers in Marine Science, 9*, Article 933111.

Triest, L., Van der Stocken, T., Allela Akinyi, A., Sierens, T., Kairo, J., & Koedam, N. (2020). Channel network structure determines genetic connectivity of landward–seaward Avicennia marina populations in a tropical bay. *Ecology and Evolution, 10*(21), 12059–12075.

Triest, L., Van der Stocken, T., Sierens, T., Deus, E. K., Mangora, M. M., & Koedam, N. (2021a). Connectivity of Avicennia marina populations within a proposed marine transboundary conservation area between Kenya and Tanzania. *Biological Conservation, 256*, Article 109040.

Triest, L., Van der Stocken, T., De Ryck, D., Kochzius, M., Lorent, S., Ngeve, M., Koedam, N., et al. (2021b). Expansion of the mangrove species Rhizophora mucronata in the Western Indian Ocean launched contrasting genetic patterns. *Scientific Reports, 11*(1), 4987.

UNCCD. (2016). *Somalia national action programme for the United Nations convention to combat desertification.* https://www.unccd.int/sites/default/files/naps/2018-06/NAP%20Full%20Report%20-%20Final%2023%20May%20digital.pdf

Vallerani, M. (2017). *Empowered Participatory Governance and marine resources management: The case of two Locally Managed Marine Areas in Southern Kenya.* https://edepot.wur.nl/441092

Warui, M. W., Manohar, S., & Obade, P. (2020). Current status, utilization, succession and zonation of mangrove ecosystem along Mida Creek, Kenya. *International Journal of Bonorowo Wetlands, 10*(1).

Waweru, B. W., Muthumbi, A. W., Vanreusel, A., Wangondu, V., & Mutua, A. (2022). Free-living marine nematode communities in Rhizophora mucronata Lam.(Rhizophoraceae) forest at Mida Creek, Kenya. *Western Indian Ocean Journal of Marine Science, 21*(2), 33–43.

5

Southern Africa's Coast: Mangroves for Resilience in the Face of Climate Change

Abstract Southern Africa's coastal ecosystems, encompassing Mozambique, South Africa, and Namibia, face growing challenges due to climate change, including rising sea levels, coastal erosion, and increased flooding. Mangrove ecosystems, which thrive in these regions, play a vital role in enhancing coastal resilience. This study explores the critical role of mangroves in mitigating climate-induced risks, particularly their ability to reduce vulnerability to storms and flooding through sediment stabilization and coastal protection. The paper highlights the diverse mangrove species in Southern Africa, such as *Rhizophora mucronata*, *Avicennia marina*, and *Bruguiera gymnorrhiza*, which contribute to the health of marine habitats and support local livelihoods. Mangroves also offer economic benefits, supporting fisheries, tourism, and coastal protection, which are increasingly threatened by climate change. Effective climate adaptation strategies, including mangrove restoration and conservation, are discussed, with an emphasis on community-based approaches and cross-border cooperation between Southern African nations. The role of mangroves in carbon sequestration is also addressed, underscoring their contribution to global climate change mitigation efforts. Collaborative efforts involving governments, NGOs, and local communities have proven successful in restoring mangrove forests, providing a model for future conservation actions. The chapter concludes with policy recommendations to integrate mangrove conservation into national and regional climate strategies, ensuring that these vital ecosystems contribute to sustainable and resilient coastal development.

Keywords Mangroves · Climate change · Coastal resilience · Southern Africa · Adaptation strategies · Carbon sequestration

5.1 Introduction to Southern Africa's Coastal Regions

Southern Africa's coastal regions are characterized by diverse ecosystems, rich biodiversity, and a variety of unique habitats (Sieben et al., 2021). Countries such as Mozambique, South Africa, Namibia, Angola, and Madagascar, which all lie along the Indian and Atlantic Oceans, boast expansive coastal areas that play critical roles in regional economies, food security, and environmental health (Garcia & Ribeiro, 2023; Mkhonto, 2022). These ecosystems, ranging from sandy beaches to mangrove forests, coral reefs, and coastal wetlands, are vital not only for local communities but also for global biodiversity. Mozambique, with its vast 2,470 kilometers of coastline along the Indian Ocean, is home to mangroves, coral reefs, and seagrass meadows, which contribute significantly to marine biodiversity and provide important resources for local fisheries (Chitará-Nhandimo et al., 2022). South Africa's coastline is equally rich, extending 2,500 kilometers and featuring dramatic cliffs, sandy shores, and sheltered coves. South Africa's marine environment is renowned for its diversity, hosting species such as the endangered great white shark and southern right whale (Cortelezzi et al., 2022; Fraser, 2022). Namibia's coastal ecosystems, although arid, sustain unique habitats, including the nutrient-rich Benguela Current that supports abundant marine life (Ruppel-Schlichting, 2022). Angola, with a coastline stretching 1,600 kilometers, is home to critical wetlands and mangrove areas that provide crucial breeding grounds for marine life (Huntley, 2023a, 2023b). Madagascar, an island nation off the coast, harbors some of the world's most unique and ecologically significant coastal ecosystems, including coral reefs, mangrove forests, and coastal rainforests (Bardou et al., 2024).

Despite their ecological importance, these coastal regions are increasingly threatened by climate change, leading to a rise in sea levels, increased flooding, and escalating coastal erosion. According to Dube et al. (2021) and Mgadle et al. (2022), Southern Africa is particularly vulnerable to climate-induced impacts, with rising sea levels projected to exacerbate coastal flooding and the erosion of key habitats. In countries like Mozambique and South Africa, the increasing frequency of severe weather events such as storms and floods disrupts coastal communities, threatening their livelihoods and causing significant economic losses (Nhundu et al., 2021; Vincent, 2024). The degradation of coastal ecosystems due to climate change has prompted growing interest in the role of mangroves in mitigating these impacts. Mangrove forests, which are found extensively along the coasts of Mozambique, Madagascar, and Angola, provide a natural buffer against the ravages of rising sea levels and coastal erosion. These ecosystems are crucial in stabilizing shorelines by trapping sediments, reducing wave energy, and preventing the loss of land to erosion (Adams & Rajkaran, 2021; Knight, 2024). Additionally, mangroves act as carbon sinks, sequestering large amounts of carbon dioxide and mitigating the impacts of climate change on a global scale (Adams et al., 2022). Mangroves also support rich biodiversity, offering nursery habitats for many fish species, and provide valuable resources to local communities, such as fuelwood, medicinal plants, and materials for construction (Sell et al., 2024). Their preservation is increasingly seen as an essential component of coastal management strategies aimed at reducing vulnerability to climate change while enhancing ecological resilience.

The coastal regions of Southern Africa are therefore of immense environmental and socio-economic value. However, they face growing threats from climate change, especially rising sea levels, flooding, and coastal erosion. The role of mangroves in mitigating these threats underscores the need for integrated coastal management strategies that prioritize the conservation of these vital ecosystems. As the region adapts to the challenges posed by climate change, strengthening the resilience of coastal ecosystems through the protection and restoration of mangroves will be a crucial step forward.

5.2 Mangrove Ecosystems for Climate Change Adaptation in Southern Africa

Mangrove ecosystems, located at the interface between land and sea, are critical to the resilience of coastal communities, particularly in regions like Southern Africa that are highly vulnerable to the impacts of climate change. These ecosystems made up of different species are essential for adaptation strategies, mitigating climate-induced risks such as rising sea levels, coastal erosion, and extreme weather events like storms and flooding (Table 5.1). Mangroves contribute significantly to both environmental and socio-economic stability, and their restoration and conservation are integral to enhancing coastal resilience. Mangroves play a vital role in reducing vulnerability to climate-induced risks by acting as natural barriers against storm surges, high winds, and flooding. According to Macamo et al. (2021), mangrove forests can reduce wave height by up to 66%, offering coastal communities critical protection. These ecosystems store large amounts of carbon, thereby helping mitigate the impacts of climate change by sequestering carbon dioxide from the atmosphere. In Southern Africa, where communities are often situated along the coast, mangroves serve as a buffer against the rising sea levels and increasingly frequent extreme weather events. The root systems of mangroves stabilize sediments, reducing coastal erosion and protecting shorelines from the detrimental effects of tidal forces and storm surges (Gullström et al., 2021).

The role of mangroves in enhancing coastal resilience is particularly pronounced in their ability to absorb and dissipate the energy of storms. Mangrove forests provide a natural defense by reducing the intensity of incoming waves and acting as physical barriers to flooding. According to Louange (2024), the presence of a healthy mangrove belt can significantly reduce the impact of tropical cyclones by dissipating wave energy and reducing storm surges. In Southern Africa, countries like Mozambique, South Africa, and Madagascar experience tropical cyclones and

Table 5.1 Mangrove species in Southern Africa

Country	Mangrove species	Description of mangrove species	References
Madagascar	*Avicennia marina* (White mangrove)	Pale bark and aerial roots that help stabilize sediments, found along the western coast, vital for sediment stabilization and marine habitat	Clausen et al. (2010), Rakotomavo and Fromard (2010), Jones et al. (2014), and Nabeelah Bibi et al. (2019)
	Rhizophora mucronata (Red mangrove)	Prop roots extending from the trunk, prevalent in Barren Isles and coastal areas, important for coastal protection and sediment trapping	Clausen et al. (2010), Rakotomavo and Fromard (2010), Jones et al. (2014), and Nabeelah Bibi et al. (2019)
	Bruguiera gymnorrhiza (Large-leafed mangrove)	Large, leathery leaves and knee-like aerial roots, found in mangrove swamps, supports wildlife and aids in sediment stabilization	Clausen et al. (2010), Rakotomavo and Fromard (2010), Jones et al. (2014), and Nabeelah Bibi et al. (2019)

(continued)

Table 5.1 (continued)

Country	Mangrove species	Description of mangrove species	References
Mozambique	*Rhizophora mucronata* (Red mangrove)	Prop roots, found along Zambezi River Delta and Quirimbas Archipelago, crucial for coastal protection and marine habitat creation	Hatton and Couto (1992), Guerreiro et al. (1996), Barbosa et al. (2001), and Penha-Lopes et al. (2009)
	Avicennia marina (White mangrove)	Salt-excreting glands, plays a key role in coastal protection and provides habitat for various species	Hatton and Couto (1992), Guerreiro et al. (1996), Barbosa et al. (2001), and Penha-Lopes et al. (2009)
	Bruguiera gymnorrhiza (Large-leafed mangrove)	Knee-like aerial roots, adapted to anoxic conditions, contributes to the structure of mangrove forests and supports marine and terrestrial organisms	Hatton and Couto (1992), Guerreiro et al. (1996), Barbosa et al. (2001), and Penha-Lopes et al. (2009)
	Ceriops tagal (Yellow mangrove)	Yellowish bark, small clustered leaves, adapted to varying salinities, contributes to biodiversity	Hatton and Couto (1992), Guerreiro et al. (1996), Barbosa et al. (2001), and Penha-Lopes et al. (2009)

(continued)

Table 5.1 (continued)

Country	Mangrove species	Description of mangrove species	References
Angola	*Rhizophora mangle* (Red mangrove)	Extensive prop roots, found in Kwanza River Delta and Luanda Bay, plays a role in coastal protection and supports marine life	Diop et al. (2002), Huntley (2019), Cardoso et al. (2021), and Huntley (2023a, 2023b)
	Avicennia marina (White mangrove)	Tolerates high salinity, helps in sediment stabilization, and is important for coastal protection and local fisheries	Diop et al. (2002), Huntley (2019), Cardoso et al. (2021), and Huntley (2023a, 2023b)
	Bruguiera gymnorrhiza (Large-leafed mangrove)	Large, leathery leaves, knee-like roots, found in Kwanza River Delta, contributes to the structure and function of mangrove ecosystems	Diop et al. (2002), Huntley (2019), Cardoso et al. (2021), and Huntley (2023a, 2023b)
South Africa	*Avicennia marina* (White mangrove)	Dominates in KwaZulu-Natal, well adapted to saline conditions, plays a crucial role in coastal protection and habitat provision	Adams et al. (2004), Hoppe-Speer et al. (2011), and Yessoufou and Stoffberg (2016)

(continued)

Table 5.1 (continued)

Country	Mangrove species	Description of mangrove species	References
	Rhizophora mucronata (Red mangrove)	Present in St. Lucia Wetland Park, prop roots help stabilize sediments and provide habitat for marine species, important for ecological balance	Adams et al. (2004), Hoppe-Speer et al. (2011), and Yessoufou and Stoffberg (2016)
	Bruguiera gymnorrhiza (Large-leafed mangrove)	Found in KwaZulu-Natal, with knee-like roots, large leaves, supports wildlife, and helps stabilize sediments in the intertidal zones	Adams et al. (2004), Hoppe-Speer et al. (2011), and Yessoufou and Stoffberg (2016)
	Ceriops tagal (Yellow mangrove)	Less common, adapted to varying salinities, contributes to biodiversity and supports local marine life	Adams et al. (2004), Hoppe-Speer et al. (2011), and Yessoufou and Stoffberg (2016)

intense rainfall events. In such regions, mangrove ecosystems act as a vital safeguard for communities by mitigating the damage caused by these storms. In addition to their role as natural barriers, mangroves also enhance coastal biodiversity. This biodiversity, in turn, supports the livelihoods of coastal populations who depend on fishing and other marine resources. The coastal ecosystem provided by mangroves is a source of food, medicine, and building materials for local communities, reinforcing the argument for integrating mangrove ecosystems into climate adaptation strategies.

Mangrove restoration has emerged as a key climate adaptation strategy in Southern Africa. The degradation of mangrove ecosystems due to human activity and climate change has exacerbated the vulnerability of coastal areas. Restoration efforts focus on the rehabilitation of degraded mangrove forests and the establishment of new mangrove plantations. According to Peacock et al. (2023), mangrove restoration is not only an environmentally sound practice but also an economically beneficial one, as healthy mangroves enhance fishery productivity, increase biodiversity, and provide carbon sequestration services. Several countries in Southern Africa have undertaken restoration initiatives. For instance, Mozambique has launched extensive mangrove restoration projects as part of its national strategy for climate adaptation. These projects involve community-based management, where local populations are engaged in the restoration process, thus ensuring long-term sustainability. Similarly, South Africa has initiated coastal management programs that emphasize the importance of conserving and restoring mangrove ecosystems to build resilience against climate change. Furthermore, the involvement of local communities in restoration activities has been shown to enhance the effectiveness of such initiatives. Mangrove restoration can also support the socio-economic development of coastal communities by providing ecosystem services, such as sustainable fisheries, ecotourism opportunities, and increased agricultural productivity (Raw et al., 2023).

Mangrove species across different Southern African countries play vital ecological roles in coastal protection, sediment stabilization, and biodiversity support (Table 5.1). In Madagascar, *Avicennia marina* (White Mangrove) is characterized by pale bark and aerial roots, stabilizing sediments and creating marine habitats (Clausen et al., 2010; Jones et al., 2014; Nabeelah Bibi et al., 2019; Rakotomavo & Fromard, 2010). *Rhizophora mucronata* (Red Mangrove) has prop roots that help trap sediments and protect the coastline, found in Barren Isles (Clausen et al., 2010; Jones et al., 2014; Nabeelah Bibi et al., 2019; Rakotomavo & Fromard, 2010). *Bruguiera gymnorrhiza* (Large-leafed Mangrove) features knee-like roots and large leaves, contributing to wildlife support and sediment stabilization (Clausen et al., 2010; Jones et al., 2014; Nabeelah Bibi et al., 2019; Rakotomavo & Fromard, 2010). In Mozambique, *Rhizophora mucronata* is found along the Zambezi River Delta, offering coastal protection (Barbosa et al., 2001; Guerreiro et al., 1996; Hatton & Couto, 1992; Penha-Lopes et al., 2009), while *Avicennia marina* helps in coastal defense and provides a habitat for various species (Barbosa et al., 2001; Guerreiro et al., 1996; Hatton & Couto, 1992; Penha-Lopes et al., 2009). *Bruguiera gymnorrhiza* contributes to the structure of mangrove forests, supporting both marine and terrestrial organisms (Barbosa et al., 2001; Guerreiro et al., 1996; Hatton & Couto, 1992; Penha-Lopes et al., 2009), and *Ceriops tagal* (Yellow Mangrove) is vital for biodiversity (Barbosa et al., 2001; Guerreiro et al., 1996; Hatton & Couto, 1992; Penha-Lopes et al., 2009). In Angola, *Rhizophora mangle* (Red Mangrove) and *Avicennia marina* both provide sediment stabilization and coastal protection (Diop et al., 2002; Cardoso et al., 2021; Huntley, 2019; Huntley, 2023a, 2023b). *Bruguiera gymnorrhiza* plays an important role in mangrove ecosystems (Diop et al., 2002; Cardoso et al., 2021; Huntley, 2019; Huntley, 2023a, 2023b). South Africa's mangroves, including *Avicennia marina, Rhizophora mucronata, Bruguiera gymnorrhiza,* and *Ceriops tagal,* also offer coastal protection, sediment stabilization, and biodiversity support (Adams et al., 2004; Hoppe-Speer et al., 2011; Yessoufou & Stoffberg, 2016).

5.3 Mangroves and Local Livelihoods in Southern African Coastal Regions

Mangroves, coastal ecosystems characterized by salt-tolerant trees, are crucial for the well-being of local communities in Southern Africa, providing diverse economic, ecological, and cultural benefits. Found in countries like Mozambique, Angola, Madagascar, and South Africa, these ecosystems are vital for local livelihoods, especially in coastal areas where fishing, tourism, and natural resource use form the backbone of community economies.

Mangroves hold significant economic value for local coastal communities in Southern Africa. One of their primary contributions is through fisheries, as they provide critical habitats for juvenile fish, shellfish, and crustaceans. These ecosystems act as nurseries for many commercially valuable species, including shrimp and fish, supporting artisanal and small-scale fisheries that are integral to local diets and income. In Mozambique, for instance, local fisheries depend heavily on mangrove ecosystems, contributing to both domestic food security and international markets (Mafuca et al., 2024). In addition to fisheries, tourism is another growing sector benefiting from mangrove ecosystems. Ecotourism, such as bird watching, kayaking, and guided tours in mangrove forests, generates significant income for communities while promoting conservation. The coastal region of Mozambique is a prime example, where mangrove forests attract tourists seeking unique natural experiences, thus supporting local businesses and creating jobs (Chibite et al., 2021). Furthermore, mangroves play a crucial role in coastal protection. Their dense root systems trap sediments and buffer the coast against waves, reducing erosion and providing natural defenses against storm surges and floods. This protective function is particularly vital for communities in low-lying areas, where infrastructure and agricultural land are highly vulnerable to extreme weather events (Whitehead, 2022). In regions like the coastal belt of Mozambique, mangroves act as natural barriers against the effects of cyclones, which have become more frequent and severe due to climate change.

Climate change poses significant threats to mangrove ecosystems and the livelihoods they support. Rising sea levels, increased storm intensity, and shifts in temperature can lead to mangrove degradation, reducing their ability to provide critical services. Coastal communities in Southern Africa are particularly vulnerable to these changes. For example, in Mozambique, rising sea levels have led to the submergence of some mangrove areas, threatening the livelihoods of communities reliant on fish and shellfish for their income and food security (Chitará-Nhandimo et al., 2022). The changing climate also exacerbates the frequency of extreme weather events, such as cyclones and floods, which further threaten coastal settlements. Mangroves' ability to reduce the impacts of these disasters is diminished when their health is compromised. As such, protecting and restoring mangrove forests becomes a crucial strategy for maintaining the resilience of coastal communities.

To combat the challenges posed by climate change and ensure the sustainability of mangrove resources, community-based strategies are critical. Local communities in Southern Africa have developed various strategies to enhance resilience, such as community-led mangrove restoration projects. These initiatives often involve the active participation of local people in planting mangrove saplings, monitoring growth, and educating others on the importance of these ecosystems for their livelihoods (Macamo, 2023; Macamo et al., 2024). Additionally, sustainable resource management practices, such as regulated harvesting of mangrove wood and non-timber products like honey and medicinal plants, can help prevent overexploitation. By establishing local resource management committees and working closely with government agencies, communities can ensure that mangrove forests are protected while still meeting their economic needs.

5.4 Collaborative Efforts for Mangrove Conservation in Southern African Coastal Regions

Mangroves are vital coastal ecosystems that provide numerous benefits, including biodiversity support, carbon sequestration, and coastal protection. In Southern Africa, these ecosystems are facing significant threats from human activities, climate change, and unsustainable development. Collaborative efforts across borders, between governments, non-governmental organizations (NGOs), and local communities, are essential for the protection and restoration of mangroves in the region.

Southern Africa's mangrove ecosystems often span multiple national borders, making cross-border cooperation crucial for their protection. Countries like Mozambique, South Africa, and Tanzania, which share mangrove habitats, have recognized the need for joint efforts in preserving these critical ecosystems. The Blue Action Fund Programme on Ecosystem-based Adaptation in the Western Indian Ocean is one such initiative that emphasizes regional collaboration (Blue Action Fund, 2023). This program aims to create integrated management strategies for coastal ecosystems, including mangroves, to address threats such as coastal erosion, habitat loss, and the effects of climate change. In addition, the Indian Ocean Rim Association (IORA), which includes several Southern African nations, promotes regional cooperation on environmental issues (IORA, 2025). IORA supports joint research initiatives, knowledge sharing, and policy development aimed at protecting mangroves and other coastal ecosystems across the Indian Ocean. By combining efforts at the regional level, these initiatives are fostering a coordinated approach to mangrove conservation, ensuring that no single nation bears the burden of protection alone.

The success of mangrove conservation in Southern Africa relies on strong partnerships between governments, NGOs, and local communities (Ravaoarinorotsihoarana et al., 2023). Governments play a key role in policy development and implementation, while NGOs provide technical expertise, funding, and advocacy. Local communities, who depend directly on mangrove ecosystems for livelihoods such as fishing and ecotourism, are often the most affected by mangrove degradation and are crucial partners in conservation efforts. One successful example of such partnerships is seen in Mozambique. The WWF Mozambique has worked alongside the Mozambican government and local communities to restore and protect the mangrove forests in the Zambezi Delta (WWF Mozambique, 2025). This collaboration has not only helped to conserve biodiversity but has also supported sustainable livelihoods through ecotourism and sustainable fisheries practices. The project has shown that when communities are involved in decision-making processes and benefit from mangrove conservation, they are more likely to adopt sustainable practices. In South Africa, the South African Environmental Observation Network (SAEON) collaborates with local communities and governmental bodies to monitor and protect mangrove ecosystems, particularly in the Eastern Cape region (SAEON, 2025). These partnerships facilitate data collection on mangrove health, which is essential for adaptive management strategies. The government's role in integrating mangrove conservation into national policies further strengthens the long-term sustainability of these efforts.

Successful mangrove restoration efforts in Southern Africa demonstrate the potential for impactful conservation initiatives. The Eden Reforestation Projects in Mozambique is one such example. The project has seen the replanting of thousands of mangrove trees in areas heavily impacted by deforestation. The initiative has also focused on educating local communities about the importance of mangroves, promoting their sustainable use while fostering community-driven management practices. In Madagascar, the Integrated Coastal Management (ICM) program has been instrumental in restoring mangrove ecosystems along the coast. Their approach combines scientific research, community participation, and policy advocacy to create comprehensive management plans. The success of this initiative lies in its focus on strengthening local capacity for sustainable mangrove management, creating a model that other nations in the region have sought to emulate.

5.5 Climate Change Mitigation Through Mangroves in Southern African Coastal Regions

Mangroves, coastal ecosystems found in the intertidal zones of tropical and subtropical regions, play a crucial role in mitigating climate change, particularly in Southern Africa. These unique forests are well regarded for their capacity to sequester and store large amounts of carbon, a process known as blue carbon storage. The importance of mangroves in climate change mitigation has garnered increasing attention from both scientists and policymakers.

Mangroves are highly effective at capturing and storing carbon dioxide, a greenhouse gas contributing to global warming. These coastal forests store carbon at significantly higher rates compared to terrestrial forests due to their waterlogged soils, which slow down the decomposition of organic matter. This results in the accumulation of carbon-rich sediments over centuries. Studies by Bacar et al. (2023) and Raw et al. (2023) have highlighted the exceptional carbon sequestration potential of mangroves, with estimates suggesting that they can store up to four times more carbon per hectare than terrestrial forests. In Southern Africa, mangrove ecosystems are critical in coastal regions like Mozambique, Madagascar, Angola, and South Africa, where they serve as both carbon sinks and protective buffers against coastal erosion and storm surges, which are exacerbated by climate change. Mangrove forests in Southern Africa contribute significantly to climate change mitigation by acting as a carbon sink (Masson, 2024). When mangrove ecosystems are disturbed or destroyed due to coastal development, aquaculture, or pollution, carbon stored within the soil is released into the atmosphere, exacerbating global warming. However, when preserved, mangroves continue to sequester carbon, helping to offset emissions from other sources. The conservation and restoration of mangrove ecosystems in Southern Africa could play an essential role in achieving national and global climate targets, such as those outlined in the Paris Agreement. Countries like Mozambique and Tanzania, with extensive mangrove habitats, are well positioned to harness this potential to reduce their carbon footprints.

To enhance the role of mangroves in climate change mitigation, it is crucial to integrate mangrove conservation into both national and international climate strategies. At the national level, Southern African countries should incorporate mangrove protection into their climate adaptation and mitigation plans, recognizing their importance in reducing vulnerability to climate change impacts, including rising sea levels and extreme weather events. For example, Mozambique's Nationally Determined Contributions (NDCs) to the Paris Agreement have already acknowledged the need for the conservation of coastal ecosystems, including mangroves, as part of broader climate resilience strategies. Internationally, Southern African nations can collaborate through regional initiatives like the Indian Ocean Rim Association (IORA) to promote mangrove restoration projects and share best practices for mangrove conservation.

To further enhance the role of mangroves in climate resilience, several policy recommendations are proposed including the strengthening of legal frameworks, supporting community-based mangrove conservation, increase funding for mangrove restoration projects, and promoting climate education and awareness. Southern African governments should enact and enforce stronger legal protections for mangrove ecosystems. This includes regulating land use along coastal zones to prevent habitat degradation due to agricultural expansion, urbanization, or industrial development. Engaging local communities in mangrove conservation efforts is critical. Incentivizing sustainable practices, such as ecotourism and sustainable fisheries, can provide alternative livelihoods while preserving the integrity of mangrove forests. International funding mechanisms, such as the Green Climate Fund, should prioritize mangrove restoration projects. Financial support can help restore degraded mangrove habitats and expand conservation efforts across the region. Governments and NGOs must raise awareness about the critical role of mangroves in climate change mitigation and encourage broader public support for conservation initiatives.

5.6 Conclusion

Southern Africa's approach to mangrove conservation reflects a growing recognition of the critical role these ecosystems play in enhancing both environmental and community resilience. Mangroves act as natural buffers against storm surges, coastal erosion, and rising sea levels, thus providing essential protection for coastal communities. Moreover, they support biodiversity, contribute to carbon sequestration, and sustain local livelihoods, particularly through fisheries and ecotourism. Southern Africa's efforts to protect and restore these vital ecosystems must continue to prioritize community involvement, ensuring that local populations are actively engaged in sustainable resource management practices. This community-based approach not only strengthens the capacity for long-term conservation but also empowers communities to adapt to climate change by diversifying income sources and building resilience. To enhance the effectiveness of these efforts, there is a pressing need for integrated, region-wide strategies for climate adaptation and mitigation. Climate change is a transboundary issue that impacts the entire Southern African region, demanding collaborative action across countries. A unified strategy can facilitate the sharing of knowledge, resources, and best practices while addressing common challenges such as habitat loss, unsustainable land use, and pollution. Regional cooperation can also streamline funding and policy support for mangrove restoration projects and ensure that efforts align with broader climate goals, including carbon emissions reduction and sustainable development. Ultimately, the future of Southern Africa's coastal ecosystems, including mangroves, depends on collective, coordinated action at both local and regional levels. By fostering partnerships between governments, NGOs, scientists, and communities, Southern Africa can build a resilient coastal zone capable of weathering the impacts of climate change, while securing a sustainable future for generations to come.

References

Adams, J. B., & Rajkaran, A. (2021). Changes in mangroves at their southernmost African distribution limit. *Estuarine, Coastal and Shelf Science, 248*, Article 107158.

Adams, J. B., Colloty, B. M., & Bate, G. C. (2004). The distribution and state of mangroves along the coast of Transkei, Eastern Cape Province, South Africa. *Wetlands Ecology and Management, 12*, 531–541.

Adams, J. B., Wasserman, J., Raw, J. L., & Van, L. (2022). Response of South African coastal wetlands to climate change. *Clim. Change Impacts Water Resourc.: Impl. Pract. Respons. Select. South Afr. Syst, 126*.

Bacar, F. F., Lisboa, S. N., & Sitoe, A. (2023). The mangrove forest of Quirimbas national park reveals high carbon stock than previously estimated in southern Africa. *Wetlands, 43*(6), 60.

Barbosa, F. M., Cuambe, C. C., & Bandeira, S. O. (2001). Status and distribution of mangroves in Mozambique. *South African Journal of Botany, 67*(3), 393–398.

Bardou, R., Friess, D. A., Gillespie, T. W., & Cavanaugh, K. C. (2024). Assessing mangrove cover change in Madagascar (1972–2019): Widespread mangrove deforestation is slowing down. *Global Ecology and Conservation, 53*, Article e03022.

Blue Action Fund. (2023). *Blue Action Fund programme on ecosystem-based adaptation in the Western Indian Ocean.* https://www.blueactionfund.org/wp-content/uploads/2023/11/EbA_Fact-Sheet.pdf

Cardoso, J. F., Costa, J. C., da Silva Neto, C., Duarte, M. C., & Monteiro-Henriques, T. (2021). *Plant communities of Namibe Saltmarshes* (Southwest of Angola).

Chibite, E. E. A., Vieira, L. R., & Morgado, F. (2021). Environmental strategies for sustainable management of mangrove forests in Mozambique. *Partnerships for the Goals, 402*–414.

Chitará-Nhandimo, S., Chissico, A., Mubai, M. E., Cabral, A. D. S., Guissamulo, A., & Bandeira, S. (2022). Seagrass invertebrate fisheries, their value chains and the role of LMMAs in sustainability of the coastal communities—Case of Southern Mozambique. *Diversity, 14*(3), 170.

Clausen, A., Rakotondrazafy, H., Ralison, H. O., Andriamanalina, A., & WWF, M. (2010). *Mangrove ecosystems in western Madagascar: an analysis of vulnerability to climate change*. WWF Study Report, 24pp.

Cortelezzi, P., Paulet, T. G., Olbers, J. M., Harris, J. M., & Bernard, A. T. (2022). Conservation benefits of a marine protected area on South African chondrichthyans. *Journal of Environmental Management, 319*, Article 115691.

Diop, E. S., Gordon, C., Semesi, A. K., Soumaré, A., Diallo, N., Guissé, A., & Ayivor, J. S., et al. (2002). Mangroves of Africa. In *Mangrove ecosystems: Function and management* (pp. 63–121). Springer Berlin Heidelberg.

Dube, K., Nhamo, G., & Chikodzi, D. (2021). Rising sea level and its implications on coastal tourism development in Cape Town, South Africa. *Journal of Outdoor Recreation and Tourism, 33*, Article 100346.

Fraser, M. (2022). Mammals of the cape of good hope nature reserve, Western Cape, South Africa. *Biodiversity Observations, 12*, 15–46.

Garcia, F. P., & Ribeiro, S. (2023). Lusophone Geopolitics: Blue Economy and Maritime Security in Contemporary Mozambique's Geopolitics. In *Portugal and the Lusophone world: Law, geopolitics and institutional cooperation* (pp. 379–400). Springer Nature Singapore.

Guerreiro, J., Freitas, S., Pereira, P., Paula, J., & Macia, A. (1996). Sediment macrobenthos of mangrove flats at Inhaca Island, Mozambique. *Cahiers De Biologie Marine, 37*(4), 309–328.

Gullström, M., Dahl, M., Lindén, O., Vorhies, F., Forsberg, S., Ismail, R. O., & Björk, M. (2021). *Coastal blue carbon stocks in Tanzania and Mozambique: support for climate adaptation and mitigation actions*. https://www.diva-portal.org/smash/get/diva2:1607255/FULLTEXT01.pdf

Hatton, J. C., & Couto, A. L. (1992). The effect of coastline changes on mangrove community structure, Portuguese Island, Mozambique. *Hydrobiologia, 247*, 49–57.

Hoppe-Speer, S. C., Adams, J. B., Rajkaran, A., & Bailey, D. (2011). The response of the red mangrove *Rhizophora mucronata* Lam. to salinity and inundation in South Africa. *Aquatic botany, 95*(2), 71–76.

Huntley, B. J. (2019). Angola in outline: physiography, climate and patterns of biodiversity. *Biodiversity of Angola: Science & conservation: A modern synthesis*, 15–42.

Huntley, B. J. (2023a). Profiles of Angola's biomes and ecoregions. *Ecology of angola: Terrestrial biomes and ecoregions* (pp. 43–68). Springer International Publishing.

Huntley, B. J. (2023b). The mangrove biome. *Ecology of angola: Terrestrial biomes and ecoregions* (pp. 383–391). Springer International Publishing.

IORA (2025). *Indian Ocean Rim Association.* https://www.iora.int/

Jones, T. G., Ratsimba, H. R., Ravaoarinorotsihoarana, L., Cripps, G., & Bey, A. (2014). Ecological variability and carbon stock estimates of mangrove ecosystems in northwestern Madagascar. *Forests, 5*(1), 177–205.

Knight, J. (2024). Nature-based solutions for coastal resilience in South Africa. *South African Geographical Journal, 106*(1), 21–50.

Louange, E. T. (2024). *Mapping the benefits of tropical coastal ecosystems: Economic value and qualitative insights for socio-economic cost-benefit analyses* (Bachelor's thesis, University of Twente). http://essay.utwente.nl/104499/1/Louange-Ella-Thesis%20Report.pdf

Macamo, C. (2023, May 30). Governance and community participation in marine and Coastal EbA in SADC. *Johannesburg: SAIIA.*

Macamo, C. D. C. F., Inácio da Costa, F., Bandeira, S., Adams, J. B., & Balidy, H. J. (2024). Mangrove community-based management in Eastern Africa: Experiences from rural Mozambique. *Frontiers in Marine Science, 11*, 1337678.

Macamo, C., Nicolau, D., Machava, V., Chitará, S., & Bandeira, S. (2021). *A contribution to Mozambique's biodiversity offsetting scheme: Framework to assess the ecological condition of mangrove forests.* BIOFUND Final Report, Mozambique.

Mafuca, J. M., Mutombene, R. J., Filipe, O., Abdula, S., Malauene, B. S., Dias, N., Roberts, M., et al. (2024). Planning for climate change resilience—Collation, update and assessment of Mozambique's marine fisheries data and management. *PloS Climate, 3*(10), Article e0000494.

Masson, L. (2024). *Unveiling green neocolonialism in Madagascar: A case study of a Malagasy company's carbon offset project in Mahajanga* (Master's thesis). https://studenttheses.uu.nl/bitstream/handle/20.500.12932/47839/LMasson_4722043_Thesis%20archive%20and%20publication.pdf?sequence=1

Mgadle, A., Dube, K., & Lekaota, L. (2022). Conservation and sustainability of coastal city tourism in the advent of seal level rise in Durban, South Africa. *Tourism in Marine Environments, 17*(3), 179–196.

Mkhonto, D. M. (2022). *An analysis of the maritime domain governance architecture in Southern Africa* (Doctoral dissertation, Stellenbosch: Stellenbosch University). https://scholar.sun.ac.za/bitstreams/5b8588c4-efa6-41f7-9518-26a5467de023/download

Nabeelah Bibi, S., Fawzi, M. M., Gokhan, Z., Rajesh, J., Nadeem, N., RR, R. K., & Pandian, S. K., et al. (2019). Ethnopharmacology, phytochemistry, and global distribution of mangroves—A comprehensive review. *Marine drugs, 17*(4), 231.

Nhundu, K., Sibanda, M., & Chaminuka, P. (2021). Economic losses from cyclones Idai and Kenneth and floods in Southern Africa: implications on Sustainable Development Goals. *Cyclones in Southern Africa: Volume 3: Implications for the Sustainable Development Goals*, 289–303.

Peacock, R., Bently, M., Rees, P., & Blignaut, J. N. (2023). The benefits of ecological restoration exceed its cost in South Africa: An evidence-based approach. *Ecosystem Services, 61*, Article 101528.

Penha-Lopes, G., Torres, P., Narciso, L., Cannicci, S., & Paula, J. (2009). Comparison of fecundity, embryo loss and fatty acid composition of mangrove crab species in sewage contaminated and pristine mangrove habitats in Mozambique. *Journal of Experimental Marine Biology and Ecology, 381*(1), 25–32.

Rakotomavo, A., & Fromard, F. (2010). Dynamics of mangrove forests in the Mangoky River delta, Madagascar, under the influence of natural and human factors. *Forest Ecology and Management, 259*(6), 1161–1169.

Ravaoarinorotsihoarana, L. A., Ratefinjanahary, I., Aina, C., Rakotomahazo, C., Glass, L., Ranivoarivelo, L., & Lavitra, T. (2023). Combining traditional ecological knowledge and scientific observations to support mangrove restoration in Madagascar. *Forests, 14*(7), 1368.

Raw, J. L., Van Niekerk, L., Chauke, O., Mbatha, H., Riddin, T., & Adams, J. B. (2023). Blue carbon sinks in South Africa and the need for restoration to enhance carbon sequestration. *Science of the Total Environment, 859*, Article 160142.

Ruppel-Schlichting, K. (2022). Namibia and its environment. In *Environmental Law and Policy in Namibia* (pp. 65–74). Nomos Verlagsgesellschaft mbH & Co. KG.

SAEON. (2025). *South African environmental observation network*. https://www.saeon.ac.za/

Sell, A. F., von Maltitz, G. P., Auel, H., Biastoch, A., Bode-Dalby, M., Brandt, P., Wilhelm, M. R., et al. (2024). Unique southern African terrestrial and oceanic biomes and their relation to steep environmental gradients. *Sustainability of Southern African ecosystems under global change: Science for management and policy interventions* (pp. 23–88). Springer International Publishing.

Sieben, E. J. J., Glen, R. P., Van Deventer, H., & Dayaram, A. (2021). The contribution of wetland flora to regional floristic diversity across a wide range of climatic conditions in southern Africa. *Biodiversity and Conservation, 30*(3), 575–596.

Vincent, J. (2024). *Climate impact resilience and community development: Adaptive solutions and challenges in rural Southern Africa–Coastal Mozambique as an example.* https://www.diva-portal.org/smash/get/diva2:1879961/FULLTEXT01.pdf

Whitehead, A. J. (2022). *Green coastal protection and flood defence options for Western Indian Ocean countries* (Doctoral dissertation, Stellenbosch: Stellenbosch University). https://scholar.sun.ac.za/bitstreams/61be54b8-67ea-4fa1-9472-42d94b31aeb9/download

WWF Mozambique. (2025). *WWF priority land and seascapes in Mozambique.* https://wwf.panda.org/wwf_offices/mozambique/

Yessoufou, K., & Stoffberg, G. H. (2016). Biogeography, threats and phylogenetic structure of mangrove forest globally and in South Africa: A review. *South African Journal of Botany, 107*, 114–120.

6

West Africa's Coast: Lessons in Resilience and Adaptation Through Sustainable Management of Mangrove Ecosystems

Abstract West Africa's coastal regions, spanning countries such as Senegal, Nigeria, and Ghana, face significant environmental pressures due to industrialization, urbanization, and oil exploration. These challenges have resulted in the degradation of crucial mangrove ecosystems, which are vital for coastal protection, biodiversity conservation, and local livelihoods. Mangroves mitigate environmental impacts such as coastal erosion and provide essential ecosystem services. However, oil spills, pollution, and urban sprawl threaten these habitats, leading to adverse socio-economic consequences for coastal communities. This study explores the resilience and adaptation lessons drawn from mangrove conservation efforts across West Africa. Case studies from countries like Nigeria, Senegal, and Guinea-Bissau highlight successful restoration projects and regional collaborations that have promoted sustainable management practices. These initiatives, supported by both local communities and policy frameworks, demonstrate the importance of integrating traditional knowledge with modern conservation techniques. Furthermore, the study emphasizes the critical role of mangroves in enhancing community resilience to climate change and contributing to long-term coastal development. Recommendations include scaling up community-led initiatives and incorporating mangrove conservation into national development plans. The findings underscore the need for a unified approach to preserve West Africa's mangrove ecosystems, ensuring their vital role in sustainable coastal management and climate change adaptation.

Keywords West Africa · Mangrove ecosystems · Coastal protection · Sustainable management · Community engagement · Climate resilience

6.1 Introduction to West Africa's Coastal Areas

West Africa's coastal areas are among the most ecologically diverse and economically significant regions of the continent (Almar et al., 2023). Stretching over 5,000 kilometers from the Atlantic Ocean, these coastal zones span several countries, including Senegal, Gambia, Sierra Leone, Nigeria, Ghana, and others. These regions are critical to the livelihoods of millions, supporting local economies through fishing, agriculture, tourism, and trade (Foli et al., 2021). The coastal areas also serve as rich habitats for biodiversity, providing valuable ecosystem services, including carbon sequestration, coastal protection, and support for fisheries. Senegal, located in the westernmost part of Africa, boasts a coastline rich in fishing resources, with its capital, Dakar, serving as a major maritime hub. The Gambia, though small, has a significant coastline along the Atlantic, heavily relying on its beaches for fishing and tourism (Dibba et al., 2025; Fanneh, 2021). Sierra Leone's coastline is marked by pristine beaches and mangrove forests, critical for biodiversity and the local economy (Feng et al., 2022). Nigeria, with one of the longest coastlines in West Africa, is home to vital ports such as Lagos and Port Harcourt, and its offshore oil reserves are globally important (Ateme, 2021; Fashae et al., 2022; Muhammed et al., 2024). Ghana, similarly, has a coastline stretching over 500 kilometers, known for its tourist destinations and fishing industry, while also being a key oil-producing nation in the region (Charuka et al., 2023; Nunoo & Agyekumhene, 2022). Beyond these countries, the entire West African coast is characterized by its unique ecosystems, including estuaries, wetlands, and rich marine life. These coastal zones are vital for food security, particularly for millions of small-scale fishermen who depend on the sea for their daily sustenance. However, they are also facing numerous environmental and socio-economic challenges.

One of the most significant challenges facing West Africa's coastal regions is the increasing exploitation of natural resources, particularly oil and gas reserves (Almar et al., 2023). Countries like Nigeria, Ghana, and Côte d'Ivoire are major players in the global oil industry, with substantial offshore drilling activities. While oil exploration has contributed to economic growth, it has also led to environmental degradation (Adeola et al., 2022). Oil spills, gas flaring, and the destruction of marine habitats have caused long-term ecological damage. In Nigeria's Niger Delta, for instance, oil pollution has severely impacted local communities, with devastating consequences for both the environment and the people who rely on it for their livelihoods (Yamou, 2024). Urbanization and industrial development also pose significant challenges to West Africa's coastal areas. Rapid urban growth in cities like Lagos, Accra, and Dakar has led to the destruction of vital ecosystems such as mangroves and wetlands, as well as increased pollution and waste (Alves et al., 2022). Additionally, urban sprawl and the expansion of industrial zones have contributed to coastal erosion, reducing the natural resilience of these regions to climate change and rising sea levels.

Mangroves, which thrive in the intertidal zones of West Africa's coastlines, play a crucial role in mitigating the environmental impacts of human activity (Aransiola et al., 2024). These unique ecosystems serve as buffers against coastal erosion, storm surges, and flooding, protecting both human settlements and biodiversity. Mangroves act as carbon sinks, sequestering significant amounts of carbon dioxide and helping to mitigate climate change (Henriques et al., 2021; Manga et al., 2022). Furthermore, they provide critical habitats for fish and other marine life, supporting the region's fishing industry. Unfortunately, mangrove forests are increasingly threatened by human activities such as urban expansion, deforestation, and oil exploration (Gnansounou et al., 2021). The loss of these ecosystems undermines the natural defenses of West Africa's coastal areas and exacerbates the vulnerability of local communities to the impacts of climate change. As such, conservation efforts aimed at preserving mangroves are critical for ensuring the long-term sustainability of West Africa's coastal regions (Sam et al., 2023).

West Africa's coastal areas particularly its mangroves are a vital resource for the region's economies, providing food, livelihoods, and ecological services. However, the rapid pace of oil exploration, urbanization, and industrial development is placing immense pressure on these ecosystems. Mangroves, in particular, are essential in mitigating the environmental impacts of these activities. Sustainable development strategies, including the protection and restoration of coastal ecosystems, are needed to ensure that West Africa's coastal areas remain resilient in the face of growing environmental and socio-economic challenges.

6.2 The Impact of Industrialization on Mangrove Ecosystems in West Africa

Industrialization in West Africa has led to significant environmental changes, with mangrove ecosystems being among the most vulnerable. These ecosystems, which are crucial for maintaining biodiversity, protecting shorelines, and supporting the livelihoods of local communities, have suffered due to the expansion of oil extraction, urban sprawl, and pollution. The consequences of these activities have far-reaching environmental, social, and economic impacts, affecting not only the health of the ecosystems but also the well-being of the communities that depend on them.

Oil spills, industrial discharge, and urbanization have been identified as key drivers of mangrove degradation in West Africa (Numbere, 2021a, 2021b). In countries like Nigeria, where oil extraction is concentrated in the Niger Delta, mangroves are routinely subjected to devastating oil spills. These spills destroy the natural habitat of mangrove forests by coating the roots of the trees with oil, depriving them of oxygen, and killing vital plant life. According to the study by Akani et al. (2022), oil contamination in the Niger Delta has led to the decimation of vast tracts of mangrove forests, exacerbating soil degradation and decreasing biodiversity. Pollution from industrial waste and urban sprawl further compounds the problem. As West African cities expand, more waste and chemicals are discharged into nearby rivers and coastal waters, contaminating mangrove environments. The chemical runoff from agricultural activities and industrial operations leads to eutrophication, which disrupts the balance of the ecosystem, promoting harmful algal blooms that suffocate mangrove roots and limit the availability of oxygen. Urbanization contributes to the direct destruction of mangrove habitats. In rapidly developing coastal areas like Lagos in Nigeria and Abidjan in Côte d'Ivoire, mangrove forests are cleared to make way for infrastructure, residential areas, and industrial parks (Ofoezie et al., 2022). This further reduces the ability of mangroves to provide essential ecological services, such as flood control, carbon sequestration, and wildlife habitat.

The destruction of mangrove ecosystems in West Africa has profound social and economic consequences (Elisha & Felix, 2021; Dada et al., 2024). Mangroves support the livelihoods of local communities through fishing, farming, and ecotourism. In the Niger Delta, for example, communities rely on mangroves for subsistence fishing and as a source of fuelwood. However, the degradation of these habitats has led to a decline in fish populations, reduced water quality, and a loss of income for many families. According to the International Union for Conservation of Nature (IUCN), the destruction of mangroves in this region has resulted in the displacement of thousands of fishermen who no longer have access to productive fishing grounds. Health implications are equally severe. Oil spills and industrial pollution release harmful chemicals, such as benzene and toluene, which contaminate local water sources and pose serious risks to human health. Chronic exposure to these toxins can lead to respiratory illnesses, skin conditions, and other long-term health issues for people living in polluted areas. Additionally, the loss of mangroves increases the vulnerability of communities to coastal flooding and storms, exacerbating the impacts of climate change.

To mitigate the environmental and social costs of industrialization on mangrove ecosystems, several measures can be taken. First, stronger enforcement of environmental regulations is critical. Governments must ensure that oil companies, industries, and urban developers adhere to environmental standards that protect mangrove habitats. Initiatives such as the restoration of degraded mangrove areas, as seen in Senegal's coastal rehabilitation programs (Arumugam et al., 2021; Macera et al., 2024), offer a promising approach to reversing some of the damage. Moreover, involving local communities in conservation efforts is essential. Collaborative management strategies that engage communities in monitoring and protecting mangrove forests can lead to more sustainable outcomes. For instance, the creation of community-based conservation programs in Ghana's coastal regions has successfully integrated local knowledge and practices to protect mangrove ecosystems.

Mangroves are widely distributed across the coasts of West Africa (Fig. 6.1). Mangrove species across West Africa vary significantly by country, each providing vital ecological services, including coastal protection, biodiversity support, and local livelihoods (Table 6.1). In Mauritania, mangroves are found predominantly in the Banc d'Arguin National Park, with species like *Avicennia germinans* and *Rhizophora mangle*. These forests are essential for marine life breeding and coastal erosion control (Dahdouh-Guebas & Koedam, 2001; Tahiri, 2023). Senegal hosts mangroves in areas like the Saloum Delta and Casamance, with species such as *Rhizophora mucronata* and *Laguncularia racemosa*, which are crucial for fisheries and bird habitats (Carré et al., 2022; Vidy, 2000). In Gambia, mangroves along the river estuary support local communities and ecosystems, featuring species like *Avicennia africana* (Bayo et al., 2022; Wetterer et al., 2023). Guinea-Bissau has large mangrove areas, particularly in the Bijagós Archipelago, home to species like *Rhizophora racemosa* and *Avicennia germinans*, which support fisheries and rich biodiversity (Dias et al., 2022; Garbanzo et al., 2024). In Guinea Conakry, species like *Rhizophora mangle* are found along the coast and river deltas, playing a key role in erosion protection (Cissoko et al., 2020; Konate et al., 2023). Sierra Leone features mangroves around the Freetown Peninsula, with species such as *Rhizophora racemosa*, crucial for fish breeding and coastal protection (Huber et al., 2023; Konoyima, 2020a, 2020b). In Liberia, mangroves in coastal wetlands, including the Sapo National Park, include *Rhizophora mangle* and *Avicennia germinans*, supporting fisheries and flood regulation (Naidoo, 2023; Olatunji & Charles, 2020). Côte d'Ivoire's mangroves, such as those near San Pedro, feature species like *Rhizophora* spp. and provide coastal protection (Egnankou et al., 2023; Mathieu & Anthelme, 2021).

Ghana's mangroves, dominated by *Rhizophora racemosa*, support fisheries and biodiversity (Adotey et al., 2022; Nunoo & Agyekumhene, 2022). Togo and Benin host mangroves in coastal estuaries, featuring species like *Avicennia africana*, important for coastal protection (Adanguidi et al., 2020; Gnansounou et al., 2022a). Nigeria boasts one of Africa's largest mangrove areas in the Niger Delta, including *Rhizophora* spp. and *Laguncularia racemosa*, vital for local fisheries and carbon sequestration (Akpovwovwo & Gbadegesin, 2022; Sam et al., 2023). Lastly, São Tomé and Príncipe has mangroves on São Tomé Island, featuring *Rhizophora mangle* and supporting biodiversity and local fisheries (Cravo, 2021; Machava-António et al., 2022).

6.3 Successful Regional Initiatives on Mangrove Conservation in West Africa and Lessons Learned

Mangrove ecosystems in West Africa provide essential services, such as coastal protection, carbon sequestration, and biodiversity preservation. Despite their importance, mangroves are under threat from deforestation, coastal development, and climate change. However, several successful mangrove conservation initiatives in the region have demonstrated the potential for restoration and sustainable management of these ecosystems. Through regional collaborations and effective policy frameworks, West Africa has made strides in protecting and restoring mangroves. Some notable examples are the Reforest'Action and Seawater Solutions (SwS) mangrove restoration projects, which aimed to restore over 500 hectares of degraded mangrove forests along the Volta River estuary particularly in the Songhor lagoon, home to the Pude community, and the Keta lagoon, home to the Fiaxor community (ReforestAction, 2025). The project which is carried out in collaboration with local communities, emphasized planting native species and engaging local people in the restoration process. The success of this project was largely due to strong community involvement and careful monitoring of the growth and health of planted mangroves. According to Nunoo

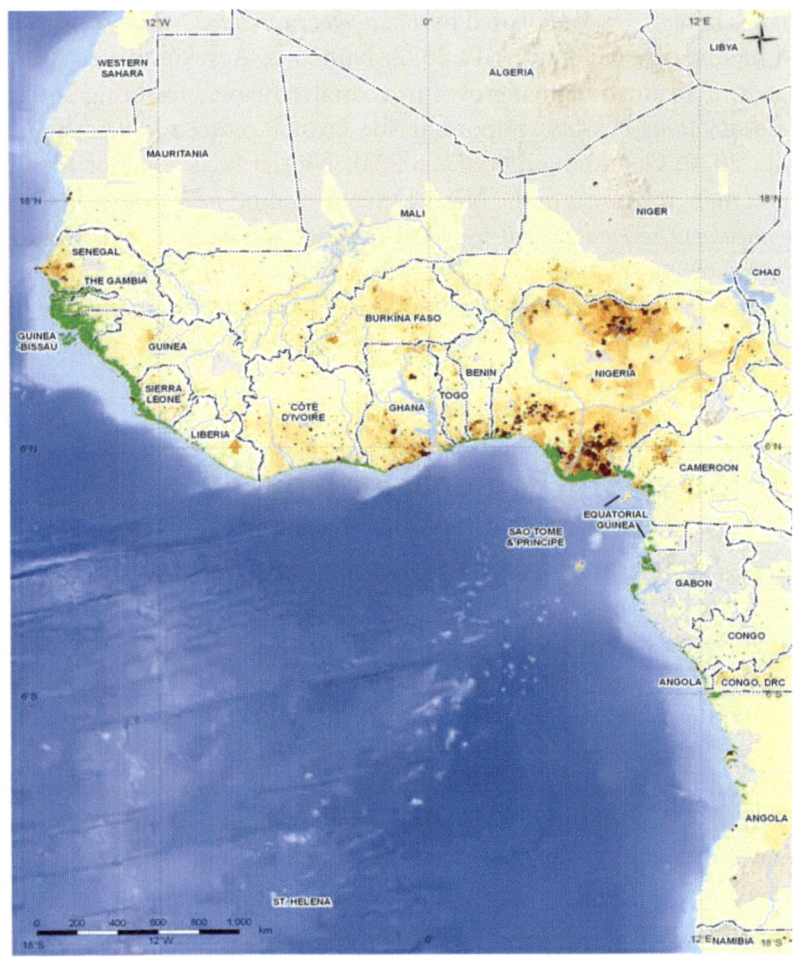

Fig. 6.1 Distribution of mangroves along the coasts of West Africa (*Source* Oyebade et al., 2010)

and Agyekumhene (2022) and Awuku-Sowah et al. (2023), community participation is crucial in ensuring the sustainability of restoration efforts, as locals take ownership of the initiative and manage the restoration sites long term. Another example is the Senegal Mangrove Restoration Project under the auspices of Livelihoods working in collaboration with the NGO Océanium, to restore over 10,000 hectares of mangrove

Table 6.1 Mangrove species diversity across countries in West Africa

Country	Mangrove species	References
Mauritania	The mangrove forests of Mauritania are mainly found along the southern coastline, especially within the Banc d'Arguin National Park, a UNESCO World Heritage site. This park contains a substantial area of mangroves, including species like *Avicennia germinans* and *Rhizophora mangle*. These mangroves are essential for marine life breeding and act as a protective barrier against coastal erosion	Dahdouh-Guebas and Koedam (2001), Otero et al. (2016), Pottier et al. (2021), Tahiri (2023), and Tahiri (2024)
Senegal	Senegal is home to vast mangrove forests, particularly in the Saloum Delta and Casamance regions. The Saloum Delta National Park stands out for its rich diversity of mangrove species, including *Rhizophora mucronata* and *Laguncularia racemosa*. These mangroves are vital for supporting local fisheries, safeguarding the coastline, and providing habitats for bird species	Vidy (2000), Conchedda et al. (2011), Bodin et al. (2013), Cormier-Salem and Panfili (2016), and Carré et al. (2022)
Gambia	The mangrove forests of The Gambia are primarily found along the river estuary and coastal regions. These mangroves, including species such as *Avicennia africana*, are vital to the livelihoods of communities reliant on fishing and agriculture. Additionally, they help mitigate coastal erosion and offer habitat for migratory birds	Bayo et al. (2022), Liman Harou et al. (2023), and Wetterer et al. (2023)

(continued)

Table 6.1 (continued)

Country	Mangrove species	References
Guinea-Bissau	Guinea-Bissau is home to some of the largest mangrove areas in West Africa, especially within the Bijagós Archipelago. These mangrove forests are composed of species like *Rhizophora racemosa* and *Avicennia germinans*. They play a crucial role in supporting local fisheries and contribute to biodiversity by providing a habitat for various fish and bird species	Dias et al. (2022), Sousa et al. (2023), and Garbanzo et al. (2024)
Guinea Conakry	Guinea's mangrove forests are primarily located along the coastline and in river deltas, such as the Konkouré River. Common species include *Rhizophora mangle* and *Avicennia africana*. These mangroves are essential for shielding coastal communities from erosion and offering a habitat for marine species	Cissoko et al. (2020), Konate et al. (2023), and Cissoko et al. (2024)
Sierra Leone	Sierra Leone's mangrove forests are found along the coast and in estuarine areas, particularly around the Freetown Peninsula and the Sierra Leone River estuary. Prominent species include *Rhizophora racemosa* and *Avicennia marina*. These mangroves are vital for providing breeding grounds for fish and safeguarding the coastline	Konoyima (2020a, 2020b), Huber et al. (2023), Suleiman and Yang (2023)
Liberia	Liberia's mangrove forests are mainly located in coastal wetlands, including areas like Sapo National Park and the Mesurado River estuary. Species such as *Rhizophora mangle* and *Avicennia germinans* are prevalent. These mangroves offer vital ecosystem services, such as flood control and support for local fisheries	Olatunji and Charles (2020), Naidoo (2023), and Kiazolu et al. (2024)

Country	Mangrove species	References
Côte d'Ivoire	Côte d'Ivoire boasts extensive mangrove forests, especially in the southwestern regions around San Pedro and Aby Lagoon. The mangrove species found here include *Rhizophora spp.*, *Avicennia spp.*, and *Laguncularia racemosa*. These mangroves are essential for coastal protection, supporting local fisheries, and providing habitat for various species	Mathieu and Anthelme (2021), Egnankou et al. (2023), Kochoni et al. (2023), and Amenoudji et al. (2024)
Ghana	In Ghana, mangroves are located along the western coast, particularly around the Volta River estuary and Keta Lagoon. Key species include *Rhizophora racemosa* and *Avicennia africana*. These mangroves play a crucial role in supporting local fisheries, controlling coastal erosion, and preserving biodiversity	Nunoo and Agyekumhene (2022), Ofori et al. (2023), Dali et al. (2023), Adotey et al. (2022), Asare and Javier (2022), Dali (2023), Dali (2020), and Aja et al. (2022)
Togo	Togo's mangrove forests are mainly found in the southeastern coastal areas, including the Togo River estuary. Common species include *Avicennia africana* and *Rhizophora mangle*. These mangroves contribute to coastal protection and provide support for local fishing communities	Gnansounou et al. (2022a, 2022b), and Gnansounou et al. (2021)
Benin	Benin's mangroves are located in the southern region, along coastal lagoons and river estuaries. Prominent species include *Rhizophora racemosa* and *Avicennia germinans*. These mangroves play a vital role in ecological functions such as protecting the coastline and providing habitat for fish and bird species	Gnansounou et al. (2021), Sinsin et al. (2022), Padonou et al. (2021), and Adanguidi et al. (2020)

(continued)

Table 6.1 (continued)

Country	Mangrove species	References
Nigeria	Nigeria is home to vast mangrove forests, especially in the Niger Delta region. These mangroves feature species such as *Rhizophora spp.*, *Avicennia spp.*, and *Laguncularia racemosa*. They are crucial for supporting local fisheries, protecting coastlines, and sequestering carbon, with the Niger Delta being one of Africa's largest mangrove areas	Akpovwovwo and Gbadegesin (2022), Sam et al. (2023), Gogo et al. (2022), Ogbeibu and Oribhabor (2023), and Numbere (2021a, 2021b)
São Tomé and Príncipe	The mangroves of São Tomé and Príncipe are primarily located along the coastlines of São Tomé Island and Príncipe Island. Key species include *Rhizophora mangle* and *Avicennia germinans*. These mangroves play a significant role in conserving biodiversity, protecting the coastline, and supporting local fishing practices	Machava-António et al. (2022), Cravo et al. (2021), Cravo (2021), and Heumüller (2021)

forests in the Sine-Saloum Delta and Casamance regions. With support from the Senegalese government and international organizations like IUCN and WWF, this project utilized both natural regeneration and active planting techniques. It highlighted the importance of involving local fishing communities, who depend on the health of the mangroves for their livelihoods. Mangrove restoration projects in Senegal also promote sustainable fisheries and ecotourism as economic alternatives to mangrove destruction (Arumugam et al., 2021; Diop et al., 2023).

Regional collaborations have been instrumental in promoting mangrove conservation in West Africa. For instance, the West Africa Marine Ecoregion (WAME) initiative, launched by IUCN in the early 2000s, focused on regional cooperation to safeguard mangroves and other critical coastal ecosystems. WAME encouraged transboundary dialogue between countries like Senegal, Guinea-Bissau, and Sierra Leone, fostering a collaborative approach to tackling shared environmental challenges. According to Osemwegie et al. (2021), regional collaboration initiatives have been successful in creating a network of protected areas, enforcing sustainable resource management practices, and developing monitoring and evaluation systems. Additionally, regional policy frameworks such as the Abidjan Convention (1976), a legally binding agreement signed by West African nations, provide a platform for the conservation of coastal and marine environments, including mangroves. It facilitates information sharing, policy coordination, and the establishment of joint environmental management programs.

Several key lessons have emerged from these successful mangrove conservation initiatives in West Africa with the most prominent being community engagement, collaborative governance, integrated approaches, as well as policy support and legal frameworks. Successful mangrove conservation projects are often those that prioritize local involvement. Active participation from local communities ensures that conservation efforts are culturally appropriate and economically beneficial, leading to long-term sustainability. Regional cooperation and collaborative governance models have proven effective in managing shared coastal resources (Kochoni et al., 2025). The WAME initiative underscores the importance of multi-country collaborations for overcoming challenges such as transboundary pollution and illegal resource extraction. Mangrove conservation is most successful when it is integrated with broader environmental, economic, and social development goals. Projects that promote alternative livelihoods, such as sustainable fisheries or ecotourism, help reduce pressure on mangrove ecosystems (Mazzuoli, 2025). Effective policy frameworks, like the Abidjan Convention, provide the necessary legal and institutional backing for mangrove protection. The enforcement of regulations, supported by robust monitoring systems, is essential for preventing illegal logging and unsustainable land use.

6.4 Sustainable Mangrove Management Practices and Community Engagement in West Africa

Mangroves are critical ecosystems that provide a range of ecosystem services, including coastal protection, biodiversity conservation, and support for livelihoods. In West Africa, mangroves are particularly vital for communities that rely on them for resources such as timber, fuelwood, and fisheries. However, mangrove ecosystems are increasingly threatened by deforestation, climate change, and unsustainable exploitation. To ensure the long-term sustainability of mangrove forests, effective management practices and community engagement are essential.

Local communities in West Africa have long depended on mangroves for various subsistence and economic activities, such as fishing, farming, and harvesting timber for building materials. As stewards of these ecosystems, they have traditionally practiced resource management strategies that are tailored to their specific environments. For example, in the coastal regions of Senegal and Guinea-Bissau, local communities often regulate the timing of resource extraction to prevent overexploitation (Sall, 2024). These traditional practices, grounded in local knowledge, have shown the potential to contribute to sustainable mangrove management, especially when they are integrated with modern conservation strategies. Recognizing the importance of local involvement in conservation efforts, numerous studies highlight the role of community-based natural resource management (CBNRM) in mangrove ecosystems. In West Africa, participatory approaches such as the case in the Volta region of Ghana have empowered communities to take ownership of mangrove restoration and management, fostering a sense of responsibility for preserving these vital ecosystems. Such engagement ensures that conservation initiatives are culturally appropriate, locally accepted, and more likely to succeed in the long term (Sam et al., 2023).

Stakeholder engagement is a key aspect of sustainable mangrove management. A variety of strategies can be employed to engage different stakeholders, including local communities, governments, NGOs, and private sector actors. One critical strategy is the development of mangrove management plans that include input from all relevant stakeholders. These plans ensure that resource use is managed equitably, with clear guidelines for sustainable harvesting and restoration activities. Education and awareness-raising campaigns are also important tools in fostering stakeholder engagement. In West Africa, initiatives like the West African subregional mangrove project have focused on educating communities about the importance of mangroves for coastal protection and livelihoods (UN, 2025). By increasing awareness of the ecological and socio-economic value of mangroves, these campaigns help build support for conservation measures and promote behavior change at the grassroots level. In addition, integrating mangrove management into broader coastal zone management frameworks is critical for ensuring that policies and actions at the national and regional levels support sustainable practices. Collaborations between governments and local communities are essential to developing and implementing policies that balance development goals with ecological sustainability (Masuda et al., 2022).

Integrating traditional ecological knowledge with modern conservation approaches is a vital aspect of sustainable mangrove management. Traditional knowledge, passed down through generations, contains valuable insights into local ecosystems and resource management practices. For instance, local communities in Sierra Leone and Liberia have long used tidal patterns to determine the best times for harvesting mangrove resources, ensuring that they do not deplete the ecosystem (EU, 2022). Modern conservation science, such as ecological restoration techniques and climate change adaptation strategies, can complement this traditional knowledge by introducing new methods for restoring degraded mangrove habitats. For example, combining mangrove replanting with traditional management practices has been shown to promote successful mangrove restoration in some areas of West Africa (Nunoo & Agyekumhene, 2022). When traditional knowledge is integrated with scientific knowledge, it not only enhances the effectiveness of conservation initiatives but also strengthens the role of local communities in managing their natural resources. This integration fosters collaboration between traditional practitioners and conservation scientists, ensuring that conservation strategies are both scientifically sound and culturally appropriate.

6.5 Future Directions for Mangrove Conservation in West Africa

Mangroves are vital coastal ecosystems that provide a wide range of ecosystem services, such as coastal protection, carbon sequestration, and support for biodiversity. However, in West Africa, mangroves are under threat from overexploitation, land-use change, and climate change, posing a significant challenge for sustainable management. To enhance the conservation of mangroves in the region, several future directions are essential: strengthening policies, scaling up community-led initiatives, and integrating mangrove conservation into national development plans.

To improve mangrove protection in West Africa, it is crucial to strengthen policies at both local and national levels. National governments need to ensure that mangrove conservation is enshrined in formal legal frameworks. According to Quenum et al. (2024), the effectiveness of mangrove protection in the region depends on the implementation of clear policies that address the drivers of mangrove degradation. Governments should adopt integrated coastal zone management strategies that promote sustainable land use while protecting mangrove forests. Additionally, West African governments could benefit from collaborating with international organizations to ensure that mangrove conservation aligns with global environmental frameworks, such as the Paris Agreement on climate change. One policy recommendation is the establishment of marine-protected areas (MPAs), ensuring that these areas are well-managed and enforced. As highlighted by Chuku et al. (2022), properly managed MPAs have been shown to enhance the resilience of coastal ecosystems, including mangroves. Furthermore, policies should address the overexploitation of mangrove resources, such as timber and firewood, through promoting sustainable harvesting practices and alternative livelihoods. Governments should also incentivize private sector involvement in mangrove conservation through tax breaks and subsidies for businesses that engage in sustainable practices.

Community-led initiatives are proving to be effective in the conservation of mangroves, and these efforts must be scaled up to maximize their impact. Communities living near mangrove ecosystems have a vested interest in their conservation, as they directly rely on them for fishing, timber, and protection from coastal erosion. According to Azumah et al. (2021), empowering local communities with knowledge and resources is essential for effective mangrove conservation. Successful initiatives, such as those led by local fishermen and women's groups, can be replicated across the region. These initiatives often combine traditional ecological knowledge with modern conservation techniques to restore degraded mangrove forests. The involvement of communities in monitoring and protecting mangroves ensures a sense of ownership and responsibility. For example, in Senegal, community-based mangrove restoration projects have led to significant improvements in forest cover (Arumugam et al., 2021). To scale up these efforts, governments and NGOs can provide training and financial support to local communities, helping them build capacity to restore and manage mangrove ecosystems sustainably.

Integrating mangrove conservation into national development plans is essential to ensure long-term sustainability. The inclusion of mangrove ecosystems in development policies can help secure funding and resources for their protection. Mangroves provide critical ecosystem services that contribute to national economies, particularly in coastal regions where tourism, agriculture, and fisheries are key industries. According to Sinsin et al. (2023), valuing ecosystem services such as carbon sequestration, fish habitat, and coastal protection can make mangroves an integral part of national economic planning. West African countries should integrate mangrove conservation into their Nationally Determined Contributions (NDCs) under the Paris Agreement. By recognizing the role of mangroves in mitigating climate change, governments can attract international funding and support. Additionally, mangrove ecosystems should be considered in the context of climate resilience and disaster risk reduction, as they provide natural buffers against storm surges and sea-level rise.

6.6 Conclusion

In West Africa, mangroves serve as natural buffers against rising sea levels, storm surges, and coastal erosion, offering critical protection to both local communities and biodiversity. These ecosystems, with their ability to adapt to varying environmental conditions, demonstrate how nature can enhance community resilience in the face of climate impacts. Through the preservation and management of mangroves sustainably, West African nations can enhance their capacity to cope with both immediate environmental challenges and long-term climate change. The critical role of mangroves in sustainable coastal development cannot be overstated. Beyond their environmental benefits, such as carbon sequestration and habitat for diverse marine species, mangroves also provide crucial socio-economic benefits to coastal communities. They support fisheries, provide timber and non-timber forest products, and promote ecotourism, which together form the foundation for sustainable livelihoods. Moreover, through ecosystem-based adaptation approaches, mangrove conservation offers a cost-effective solution to climate-related challenges, promoting both ecological and economic sustainability. West Africa's experiences with mangrove management highlight the importance of integrating community knowledge and practices with scientific research and policy frameworks. Successful mangrove restoration and protection initiatives have demonstrated the effectiveness of local involvement and cross-sectoral collaboration, ensuring the long-term viability of these ecosystems. Ultimately, fostering resilient and adaptive coastal communities through the sustainable management of mangroves will contribute to the broader goals of climate adaptation, biodiversity conservation, and sustainable development in West Africa. Through the continuous prioritization of mangrove protection and restoration, West Africa can chart a path toward a more resilient and sustainable coastal future.

References

Adanguidi, J., Padonou, E. A., Zannou, A., Houngbo, S. B., Saliou, I. O., & Agbahoungba, S. (2020). Fuelwood consumption and supply strategies in mangrove forests-Insights from RAMSAR sites in Benin. *Forest Policy and Economics, 116*, Article 102192.

Adeola, A. O., Akingboye, A. S., Ore, O. T., Oluwajana, O. A., Adewole, A. H., Olawade, D. B., & Ogunyele, A. C. (2022). Crude oil exploration in Africa: Socio-economic implications, environmental impacts, and mitigation strategies. *Environment Systems and Decisions, 42*(1), 26–50.

Adotey, J., Acheampong, E., Aheto, D. W., & Blay, J. (2022). Carbon stocks assessment in a disturbed and undisturbed mangrove forest in Ghana. *Sustainability, 14*(19), 12782.

Aja, D., Miyittah, M. K., & Angnuureng, D. B. (2022). Quantifying mangrove extent using a combination of optical and radar images in a wetland complex, Western region, Ghana. *Sustainability, 14*(24), 16687.

Akani, G. C., Amuzie, C. C., Alawa, G. N., Nioking, A., & Belema, R. (2022). Factors militating against biodiversity conservation in the Niger Delta, Nigeria: The way out. In *Biodiversity in Africa: Potentials, threats and conservation* (pp. 573–600). Springer Nature Singapore.

Akpovwovwo, U. E., & Gbadegesin, A. (2022). Species composition and distribution patterns of the Mangrove forests of the Western Niger Delta, Nigeria. *African Geographical Review, 41*(4), 468–482.

Almar, R., Stieglitz, T., Addo, K. A., Ba, K., Ondoa, G. A., Bergsma, E. W., Arino, O., et al. (2023). Coastal zone changes in West Africa: Challenges and opportunities for satellite earth observations. *Surveys in Geophysics, 44*(1), 249–275.

Alves, R. B., Bapentire, A. D., Almar, R., Louarn, A., Rossi, P. L., Corsini, L., & Morand, P. (2022). Compendium: Coastal management practices in West Africa: existing and potential solutions to control coastal erosion, prevent flooding and mitigate damage to society. *The World Bank*.

Amenoudji, C. I., Sanogo, S., Djogli, K. R., Adanve, J. F., & Sodedji, K. F. A. (2024). Effect of salinity and substrate on the emergence and growth of propagules of the mangrove species Rhizophora racemosa in the Sassandra-Dagbego Ramsar Complex, Côte d'Ivoire. *Annual Research & Review in Biology, 39*(7), 21–31.

Aransiola, S. A., Zobeashia, S. L. T., Ikhumetse, A. A., Musa, O. I., Abioye, O. P., Ijah, U. J. J., & Maddela, N. R. (2024). Niger Delta mangrove ecosystem: Biodiversity, past and present pollution, threat and mitigation. *Regional Studies in Marine Science, 103568*.

Arumugam, M., Niyomugabo, R., Dahdouh-Guebas, F., & Hugé, J. (2021). The perceptions of stakeholders on current management of mangroves in the Sine-Saloum Delta, Senegal. *Estuarine, Coastal and Shelf Science, 248*, Article 107160.

Asare, N. K., & Javier, J. L. (2022). Tidal influence on fish faunal occurrence and distribution in an estuarine mangrove system in Ghana. *African Journal of Aquatic Science, 47*(1), 88–99.

Ateme, M. E. (2021). Developing marine and coastal resources in Nigeria: Prospects and challenges. *Maritime Technology and Research, 3*(4), 335–347.

Awuku-Sowah, E. M., Graham, N. A., & Watson, N. M. (2023). The contributions of mangroves to physiological health in Ghana: Insights from a qualitative study of key informants. *Wellbeing, Space and Society, 4*, Article 100137.

Azumah, D. M. Y., Foli, B. A. K., Williams, I. K., Agyekum, K. A., Boakye, A. A., & Wiafe, G. (2021). Capacity strengthening towards application of earth observation tools and services to enhancing marine and coastal areas management in west Africa. *Remote Sensing in Earth Systems Sciences*, 1–13.

Bayo, B., Habib, W., & Mahmood, S. (2022). Spatio-temporal assessment of mangrove cover in the Gambia using combined mangrove recognition index. *Advanced Remote Sensing, 2*(2), 74–84.

Bodin, N., N'Gom-Kâ, R., Kâ, S., Thiaw, O. T., De Morais, L. T., Le Loc'h, F., Chiffoleau, J. F., et al. (2013). Assessment of trace metal contamination in mangrove ecosystems from Senegal, West Africa. *Chemosphere, 90*(2), 150–157.

Carré, M., Quichaud, L., Camara, A., Azzoug, M., Cheddadi, R., Ochoa, D., Thomas, Y., et al. (2022). Climate change, migrations, and the peopling of sine-Saloum mangroves (Senegal) in the past 6000 years. *Quaternary Science Reviews, 293*, Article 107688.

Charuka, B., Angnuureng, D. B., & Agblorti, S. K. (2023). Mapping and assessment of coastal infrastructure for adaptation to coastal erosion along the coast of Ghana. *Anthropocene Coasts, 6*(1), 11.

Chuku, E. O., Effah, E., Adotey, J., Abrokwah, S., Adade, R., Okyere, I., Crawford, B., et al. (2022). Spotlighting women-led fisheries livelihoods toward sustainable coastal governance: The estuarine and mangrove ecosystem shellfisheries of West Africa. *Frontiers in Marine Science, 9*, Article 884715.

Cissoko, B., Camara, A., Kante, C., & Sakouvogui, A. (2020). Impacts Of Rice and Salt Growing on Mangroves in The Dubréka Area. *International Journal of All Research Writings, 2*(1), 130–136.

Cissoko, B., Kante, C., Camara, A., & Sakouvogui, A. (2024). *Impact of logging and fish smoking on mangroves in management units 5 and 7 in Sangareya-Dubréka (Guinea)*.

Conchedda, G., Lambin, E. F., & Mayaux, P. (2011). Between land and sea: Livelihoods and environmental changes in mangrove ecosystems of Senegal. *Annals of the Association of American Geographers, 101*(6), 1259–1284.

Cormier-Salem, M. C., & Panfili, J. (2016). Mangrove reforestation: Greening or grabbing coastal zones and deltas? Case studies in Senegal. *African Journal of Aquatic Science, 41*(1), 89–98.

Cravo, M. M. (2021). *Fish assemblages at Praia Salgada mangrove, Príncipe Island (Gulf of Guinea)* (Doctoral dissertation).

Cravo, M., Almeida, A. J., Lima, H., Azevedo e Silva, J., Bandeira, S., Machava-António, V., & Paula, J. (2021). Fish assemblages in a small mangrove system on Príncipe Island, Gulf of Guinea. *Frontiers in Marine Science, 8*, 721692.

Dada, O. A., Almar, R., & Morand, P. (2024). Coastal vulnerability assessment of the West African coast to flooding and erosion. *Scientific Reports, 14*(1), 890.

Dahdouh-Guebas, F., & Koedam, N. (2001). Are the northernmost mangroves of West Africa viable?—A case study in Banc d'Arguin National Park, Mauritania. *Hydrobiologia, 458*, 241–253.

Dali, G. L. A. (2020). *Assessment of the ecological health of mangrove forests along the Kakum and Pra estuaries in Ghana* (Doctoral dissertation, University of Cape Coast).

Dali, G. L. A. (2023). Litter production in two mangrove forests along the coast of Ghana. *Heliyon, 9*(6).

Dali, G. L., Aheto, D. W., & Blay, J. (2023). Mangrove resource utilization and impacts in the Pra and Kakum estuaries of Ghana. *Regional Studies in Marine Science, 63*, Article 103035.

Dias, G. A., Vasconcelos, M. J., & Catarino, L. (2022). Examining the socioeconomic benefits of oysters: A provisioning ecosystem service from the mangroves of Guinea-Bissau, West Africa. *Journal of Coastal Research, 38*(2), 355–360.

Dibba, B., Yaffa, S., Sawaneh, M., & Adzawla, W. (2025). Land cover transformation and population growth: Impacts on coastal environment of the Gambia (1990–2020). *Sustainability, 17*(5), 1853.

Diop, S. M., Thiam, M., Ndiaye, O., Ndiaye, S., & Cisse, C. (2023). Comparative and prospective evaluation of the carbon potential of the mangrove of the Sine-Saloum Delta (Senegal) from 2016 to 2021. *American Journal of Plant Sciences, 14*(9), 994–1008.

Egnankou, M. W., Gnagbo, A., Pagny, J. P. F., & Tiebre, M. S. (2023). Gestion durable des mangroves du complexe lagunaire de Grand-Lahou (Côte d'Ivoire) dans un contexte de pressions anthropiques: Sustainable management of mangroves in the Grand-Lahou lagoon complex (Côte d'Ivoire) in a context of anthropic pressures. *International Journal of Biological and Chemical Sciences, 17*(2), 505–518.

Elisha, O. D., & Felix, M. J. (2021). Destruction of coastal ecosystems and the vicious cycle of poverty in Niger Delta Region. *Journal of Global Agriculture and Ecology, 11*(2), 7–24.

EU. (2022). *A Preliminary Assessment of Ecosystem Services in the Sherbro River Estuary, Southern Sierra Leone*, 44p. https://www.eeas.europa.eu/sites/default/files/documents/SRE_Final%20Report_Ecosystem_Services_Assessment.pdf

Fanneh, M. M. (2021). Socioeconomic study of climate change and its impacts on livelihoods of people living around the coastal areas of the Gambia. *Journal of Accounting, Business and Finance Research, 13*(1), 26–36.

Fashae, O. A., Tijani, M. N., Adekoya, A. E., Tijani, S. A., Adagbasa, E. G., & Aladejana, J. A. (2022). Comparative assessment of the changing pattern of land cover along the southwestern coast of Nigeria using GIS and Remote Sensing techniques. *Scientific African, 17*, Article e01286.

Feng, C., Huang, H., Qu, T., Fan, R., Coker, I. C., Seisay, L. D., Li, L., et al. (2022). Temporal and spatial patterns of demersal fish assemblages in the coastal water of Sierra Leone. *Regional Studies in Marine Science, 56*, Article 102674.

Foli, B. A. K., Williams, I. K., Boakye, A. A., Azumah, D. M. Y., Agyekum, K. A., & Wiafe, G. (2021). Earth observation services in support of West Africa's blue economy: Coastal resilience and climate impacts. *Remote Sensing in Earth Systems Sciences*, 1–12.

Garbanzo, G., Cameira, M. D. R., & Paredes, P. (2024). The Mangrove swamp rice production system of Guinea Bissau: Identification of the main constraints associated with soil salinity and rainfall variability. *Agronomy, 14*(3), 468.

Gnansounou, S. C., Salako, K. V., Sagoe, A. A., Mattah, P. A. D., Aheto, D. W., & Glèlè Kakaï, R. (2022a). Mangrove ecosystem services, associated threats and implications for wellbeing in the Mono Transboundary Biosphere Reserve (Togo-Benin), West-Africa. *Sustainability, 14*(4), 2438.

Gnansounou, S. C., Sagoe, A. A., Mattah, P. A. D., Salako, K. V., Aheto, D. W., & Glèlè Kakaï, R. (2022b). The co-management approach has positive impacts on mangrove conservation: evidence from the mono transboundary biosphere reserve (Togo-Benin), West Africa. *Wetlands Ecology and Management, 30*(6), 1245–1259.

Gnansounou, S. C., Toyi, M., Salako, K. V., Ahossou, D. O., Akpona, T. J. D., Gbedomon, R. C., Kakaï, R. G., et al. (2021). Local uses of mangroves and perceived impacts of their degradation in Grand-Popo municipality, a hotspot of mangroves in Benin, West Africa. *Trees, Forests and People, 4*, Article 100080.

Gogo, T. E. A., Ukoima, H. N., & Chukunda, F. A. (2022). Floristic abundance and diversity of mangrove in polluted soil of Ikuru Town, Andoni, Rivers State, Nigeria. *International Journal of Agriculture and Earth Science, 8*(3), 17–28.

Henriques, M., Granadeiro, J. P., Piersma, T., Leão, S., Pontes, S., & Catry, T. (2021). Assessing the contribution of mangrove carbon and of other basal sources to intertidal flats adjacent to one of the largest West African mangrove forests. *Marine Environmental Research, 169*, Article 105331.

Heumüller, J. A. (2021). *Fish diversity in mangroves of São Tomé Island (Gulf of Guinea)* (Doctoral dissertation).

Huber, L. C., Sainge, M. N., Feka, Z. N., Kamara, R. A., Kamara, A., Sullivan, M., & Cuni-Sanchez, A. (2023). Human-driven degradation impacts on mangroves in southern Sierra Leone. *Trees, Forests and People, 14*, Article 100445.

Kiazolu, O. G., Mwaura, F., & Thenya, T. (2024). *Assessment of level of public knowledge towards mangrove forest conservation.* A Case Study of Mesurado Wetland in Liberia.

Kochoni, B. I., Avakoudjo, H. G. G., Kamelan, T. M., Sinsin, C. B. L., & Kouamelan, E. P. (2023). Contribution of mangroves ecosystems to coastal communities' resilience towards climate change: A case study in southern Cote d'Ivoire. *GeoJournal, 88*(4), 3935–3951.

Kochoni, B. I., Salako, K. V., Danquah, J. A., Sinsin, C. B. L., Mensah, S., & Glèlè Kakaï, R. (2025). Contribution of the Ramsar convention to the conservation of West-African mangroves: A case study in Benin. *Wetlands Ecology and Management, 33*(1), 1–21.

Konate, D., Bangoura, Y., Camara, O. I., & Bangoura, S. A. (2023). Human impacts on the coastal ecosystem of Tabounsou and sustainable management measures case of Matoto-Conakry. *American Journal of Environmental Science and Engineering, 9*(4), 74–81.

Konoyima, K. J. (2020a). Mangrove ecosystem resources: Dependence of coastal communities in the Scarcies River Estuary, Sierra Leone. *International Journal of Community Research, 9*(1), 2–12.

Konoyima, K. J. (2020b). Conservation of Mangroves: Challenges and prospects in the Scarcies River Estuary, Sierra Leone. *Journal of Development and Communication Studies (Indexed: African Journals Online)(In Review)*.

Liman Harou, I., Inyele, J., Minang, P., & Duguma, L. (2023). Understanding the states and dynamics of mangrove forests in land cover transitions of the Gambia using a Fourier transformation of Landsat and MODIS time series in Google Earth Engine. *Frontiers in Forests and Global Change, 5*, Article 934019.

Macera, L., Deslarzes, K., Crook, O. J., Pioch, S., & Andrieu, J. (2024). Monitoring mangrove restoration projects in Senegal, Benin, Costa-Rica, and Philippines using remote sensing. *Dynamiques environnementales. Journal international de géosciences et de l'environnement* (53).

Machava-António, V., Fernando, A., Cravo, M., Massingue, M., Lima, H., Macamo, C., Paula, J., et al. (2022). A comparison of mangrove forest structure and ecosystem services in Maputo Bay (Eastern Africa) and Príncipe Island (Western Africa). *Forests, 13*(9), 1466.

Manga, B. A. B., Diatta, A. A., Ndour, N., & Dasylva, M. (2022). Assessment of carbon sequestration by mangrove plantations in Casamance (Oussouye, Ziguinchor, Senegal). https://rivieresdusud.uasz.sn/bitstream/handle/123456789/1943/manga_article_2022.pdf?sequence=1&isAllowed=y

Masuda, H., Kawakubo, S., Okitasari, M., & Morita, K. (2022). Exploring the role of local governments as intermediaries to facilitate partnerships for the sustainable development goals. *Sustainable Cities and Society, 82*, Article 103883.

Mathieu, E. W., & Anthelme, G. (2021). Status and perspectives of mangrove management in Côte d'Ivoire. *GSC Advanced Research and Reviews, 9*(2), 045–050.

Mazzuoli, S. (2025). *Community based ecotourism for biodiversity in Africa.* https://stud.epsilon.slu.se/20794/1/mazzuoli-s-20250127.pdf

Muhammed, A., Aminu, B. M., Musa, I. O., Abdulsalam, M., Isma'il, R., Gimba, Y. M., & Moses, E. O., et al. (2024). Blue economy in Nigeria and the African Continent. In *Marine Bioprospecting for Sustainable Blue-bioeconomy* (pp. 355–370). Springer Nature Switzerland.

Naidoo, G. (2023). The mangroves of Africa: A review. *Marine Pollution Bulletin, 190*, Article 114859.

Numbere, A. O. (2021a). Impact of Urbanization and Crude oil exploration in Niger delta mangrove ecosystem and its livelihood opportunities: a footprint perspective. *Agroecological Footprints Management for Sustainable Food System*, 309–344.

Numbere, A. O. (2021b). Natural seedling recruitment and regeneration in deforested and sand-filled Mangrove forest at Eagle Island, Niger Delta, Nigeria. *Ecology and Evolution, 11*(7), 3148–3158.

Nunoo, F. K., & Agyekumhene, A. (2022). Mangrove degradation and management practices along the coast of Ghana. *Agricultural Sciences, 13*(10), 1057–1079.

Ofoezie, E. I., Eludoyin, A. O., Udeh, E. B., Onanuga, M. Y., Salami, O. O., & Adebayo, A. A. (2022). Climate, urbanization and environmental pollution in West Africa. *Sustainability, 14*(23), 15602.

Ofori, S. A., Asante, F., Boateng, T. A. B., & Dahdouh-Guebas, F. (2023). The composition, distribution, and socio-economic dimensions of Ghana's mangrove ecosystems. *Journal of Environmental Management, 345*, Article 118622.

Ogbeibu, A. E., & Oribhabor, B. J. (2023). The Niger Delta Mangrove ecosystem and its conservation challenges. In *Mangrove biology, ecosystem, and conservation.* IntechOpen.

Olatunji, E. T., & Charles, J. (2020). Change detection analysis of mangrove ecosystems in the Mesurado Wetland, Montserrado County, Liberia. *International Journal of Research in Environmental Studies, 7*, 17–24.

Osemwegie, I., Delgado, K. D. C., Arimiyaw, A. W., Kanneh, A. B., Todota, C. T., Faye, A., & Akinyemi, F. O. (2021). Diagnostic analysis of the Canary Current System of West Africa: The need for a paradigm shift to proactive natural resource management. *Ocean and Coastal Research, 69*(suppl 1), Article e21043.

Otero, V., Quisthoudt, K., Koedam, N., & Dahdouh-Guebas, F. (2016). Mangroves at their limits: Detection and area estimation of mangroves along the Sahara Desert Coast. *Remote Sensing, 8*(6), 512.

Oyebade, B. A., Emerhi, E. A., & Ekeke, B. A. (2010). Quantitative review and distribution status of mangrove forest species in West Africa. *African Research Review, 4*(2).

Padonou, E. A., Gbaï, N. I., Kolawolé, M. A., Idohou, R., & Toyi, M. (2021). How far are mangrove ecosystems in Benin (West Africa) conserved by the Ramsar Convention? *Land Use Policy, 108*, Article 105583.

Pottier, A., Catry, T., Trégarot, E., Maréchal, J. P., Fayad, V., David, G., Failler, P., et al. (2021). Mapping coastal marine ecosystems of the National Park of Banc d'Arguin (PNBA) in Mauritania using Sentinel-2 imagery. *International Journal of Applied Earth Observation and Geoinformation, 102*, Article 102419.

Quenum, I. A., Avocèvou-Ayisso, C., Idohou, R., Padonou, E. A., Akabassi, G. C., & Akakpo, B. A. (2024). Restoration and governance approaches of Mangrove ecosystems in Africa. *Wetlands, 44*(5), 54.

ReforestAction. (2025). *Volta Delta—Songhor region.* Mangrove Restoration, Ghana. https://www.reforestaction.com/en/projects/1747

Sall, N. (2024). *Overfishing in Senegal: A deep dive into the livelihood of coastal communities* (Doctoral dissertation, Université d'Ottawa/University of Ottawa). https://ruor.uottawa.ca/bitstreams/902a708e-bfc1-4b17-94e0-3ed7e817865e/download

Sam, K., Zabbey, N., Gbaa, N. D., Ezurike, J. C., & Okoro, C. M. (2023). Towards a framework for mangrove restoration and conservation in Nigeria. *Regional Studies in Marine Science, 66*, Article 103154.

Sinsin, C. B. L., Bonou, A., Salako, K. V., Gbedomon, R. C., & Glèlè Kakaï, R. L. (2023). Economic valuation of mangroves and a linear mixed model-assisted framework for identifying its main drivers: A case study in Benin. *Land, 12*(5), 1094.

Sinsin, C. B. L., Salako, K. V., Fandohan, A. B., Kouassi, K. E., Sinsin, B. A., & Glèlè Kakaï, R. (2022). Potential climate change induced modifications in mangrove ecosystems: A case study in Benin, West Africa. *Environment, Development and Sustainability, 24*(4), 4901–4917.

Sousa, J., Campos, R., Mendes, O., Duarte Lopes, P., Matias, M., Rosa, A. P., Catarino, L., et al. (2023). The (dis) engagement of mangrove forests and mangrove rice in academic and non-academic literature on Guinea-Bissau–a systematic review protocol. *PLoS ONE, 18*(4), Article e0284266.

Suleiman, A. K., & Yang, C. (2023). *Urbanization and Wetland degradation: Land use land cover change analysis of the Aberdeen Creek.* Sierra Leone River Estuary.

Tahiri, A. Z. (2023). *A successful approach to mangrove ecosystem restoration in the Mauritanian side of the Senegal River Delta.*

Tahiri, A. Z. (2024). A quest for desert forests from Eritrean to Mauritanian Mangroves. *Ecological Restoration, 42*(3), 161–168.

UN. (2025). *West African subregional mangrove project.* https://sdgs.un.org/partnerships/west-african-subregional-mangrove-project

Vidy, G. (2000). Estuarine and mangrove systems and the nursery concept: Which is which? The case of the Sine Saloum system (Senegal). *Wetlands Ecology and Management, 8*(1), 37–51.

Wetterer, J. K., Gómez, K., Keita, M., & Jallow, M. (2023). Ants nesting in red mangroves of The Gambia (West Africa). *Belgian Journal of Entomology, 143,* 1–8.

Yamou, T. A. (2024). *Addressing environmental damage to enhance agricultural livelihoods and improve the well-being of local communities: The case of crude oil pollution in Nigeria* (Doctoral dissertation, University of Kent).

7

Conclusion

Abstract Mangrove ecosystems play an essential role in building climate resilience along Africa's coasts, offering critical protection against natural disasters, coastal erosion, and flooding. This study underscores the multifaceted benefits of mangroves, which not only enhance environmental resilience but also provide significant socio-economic support through fisheries, ecotourism, and carbon sequestration. Despite the threats posed by climate change, pollution, and overexploitation, mangroves remain a vital natural resource for millions of coastal residents. The success of community-driven conservation initiatives in regions such as the Red Sea and West Africa highlights the importance of local knowledge and participation in safeguarding these ecosystems. This book advocates for the integration of mangrove conservation into national and regional climate adaptation strategies, emphasizing policy development, cross-sectoral partnerships, and sustainable management practices. Furthermore, it calls for greater investment in research, education, and capacity-building to strengthen the long-term resilience of mangrove ecosystems. Ultimately, the study emphasizes the need for collective action to protect these vital ecosystems, ensuring the socio-economic stability and survival of Africa's coastal communities amidst growing climate challenges.

Keywords Mangroves · Climate resilience · Coastal protection · Community conservation · Socio-economic benefits · Carbon sequestration · Sustainable management

7.1 Key Insights

Mangrove ecosystems play a primordial role in building climate resilience along Africa's coasts. As the impacts of climate change intensify, it is increasingly clear that mangroves are not only valuable for their biodiversity but also for their ability to protect coastal communities from the ravages of natural disasters. These ecosystems serve as natural buffers against storms, flooding, and coastal erosion, while simultaneously sustaining local livelihoods through fisheries, ecotourism, and carbon sequestration. Through a synthesis of the findings presented, it becomes evident that mangroves offer multiple services that are integral to the survival of coastal communities, ecosystems, and economies across the continent.

Africa's coastal regions are characterized by their diversity, each facing unique challenges in the face of climate-induced threats and industrial pressures. From rising sea levels to increasing storm frequency, the threats to coastal stability are varied. Furthermore, overexploitation, pollution, and urbanization place additional stress on these delicate ecosystems. Despite these challenges, mangroves persist as essential components of coastal resilience, and their conservation is pivotal for long-term sustainability.

The role of community-driven solutions and the importance of local knowledge in mangrove management have also emerged as central themes. The capacity of local populations to engage in sustainable practices, often guided by centuries-old traditions, has proven to be a powerful tool in mangrove conservation. By engaging communities in the process, conservation efforts become more effective and ensure that local needs and knowledge are incorporated into broader strategies for ecosystem management.

7.2 The Vital Role of Mangroves to Climate Resilience Along Africa's Coasts

The multifaceted role of mangroves in fostering both environmental and community resilience cannot be overstated. Mangrove ecosystems function as a vital barrier, reducing the impact of storm surges, controlling coastal erosion, and offering protection against flooding. These unique trees, which grow in saline, brackish environments, have a dense root system that stabilizes the soil and absorbs the energy of tidal waves and storm surges. By doing so, they protect coastal infrastructure, homes, and other valuable assets, safeguarding lives and livelihoods.

Furthermore, mangroves provide a critical habitat for diverse marine life, supporting fisheries that are essential to local economies. The sheltered waters within mangrove forests are nurseries for many fish species, contributing significantly to both commercial and subsistence fishing industries. In addition to supporting marine biodiversity, mangroves are also integral in sustaining ecotourism, which has become an increasingly important economic driver in many African coastal regions.

Mangroves also play a critical role in carbon sequestration, acting as powerful "carbon sinks." This function helps mitigate the effects of climate change, absorbing and storing significant amounts of carbon dioxide from the atmosphere. As the world confronts the challenges of climate change, the carbon storage capacity of mangroves positions them as key players in global efforts to combat climate warming.

The role of mangroves extends beyond environmental protection. They provide invaluable social and economic benefits to communities living along Africa's coasts. Through their ability to provide resources like firewood, construction materials, and medicinal plants, mangroves have long been integral to local economies and ways of life. Thus, their conservation is not merely an environmental concern; it is a socio-economic imperative.

7.3 Lessons from Successful Case Studies Along the Coast of Africa

Across Africa, numerous successful case studies highlight the effectiveness of community-driven approaches to mangrove conservation and restoration. These case studies emphasize the importance of local participation, collaboration, and the integration of traditional knowledge into conservation practices. In the Red Sea, for instance, community-led restoration projects have revitalized degraded mangrove forests, bringing tangible benefits to local communities in terms of food security and economic stability. In West Africa, innovative mangrove management practices have empowered coastal communities to actively engage in the protection and restoration of mangrove ecosystems, fostering both environmental sustainability and economic development.

These examples underscore the critical lesson that community involvement is a key driver of successful conservation efforts. When local communities are empowered and given ownership over the management of their natural resources, they are more likely to invest in long-term sustainable practices. Additionally, the integration of traditional ecological knowledge with modern conservation science can provide more effective and culturally appropriate solutions. By learning from these successful case studies, it becomes clear that local solutions can be scaled up and adapted to suit the unique contexts of different African coastal regions.

7.4 The Socio-economic Value of Mangroves along Africa's Coasts

The economic value of mangroves has often been underestimated, yet it is clear that these ecosystems are integral to the livelihoods of millions of people across Africa's coastal regions. Beyond their role in protecting against environmental threats, mangroves contribute significantly to the economy through fisheries, ecotourism, and the sustainable use of natural resources. Mangrove forests support a diverse range of fish species, providing the foundation for artisanal and commercial fisheries that are vital to food security and employment in coastal communities.

In addition, mangrove ecosystems are an important source of income through ecotourism. Mangrove forests attract tourists seeking unique natural experiences, whether through bird watching, boat tours, or simply enjoying the beauty and tranquility of these environments. This burgeoning sector provides economic opportunities while also promoting environmental awareness and education.

Furthermore, mangroves are a source of various non-timber forest products, such as firewood, medicinal plants, and materials for construction. These resources provide a buffer against poverty, especially in rural coastal areas where access to other economic opportunities may be limited. By quantifying the economic benefits of mangrove conservation, this book highlights the necessity of incorporating the true value of mangrove ecosystems into policy discussions and development strategies.

7.5 Collaborative Approaches to Sustainable Management of Mangroves Along Africa's Coasts

As the challenges facing Africa's mangrove ecosystems grow more complex, it is increasingly clear that a collaborative approach to conservation is essential. No single entity, whether government, non-governmental organization (NGO), or community, can tackle the magnitude of the issues on its own. Cross-sectoral partnerships that bring together governments, NGOs, local communities, private sector actors, and international organizations are key to enhancing mangrove conservation and climate adaptation efforts.

The role of policy development in these collaborative efforts is critical. Effective governance is necessary to establish clear guidelines for mangrove protection, restoration, and sustainable use. Furthermore, policies that integrate environmental concerns into national and regional development plans are essential for long-term success. Regional cooperation and capacity-building initiatives can also help strengthen the management of mangrove ecosystems, ensuring that resources are allocated efficiently and that best practices are shared across borders.

Collaborative efforts also require capacity-building, both at the community level and within governmental and non-governmental organizations. By investing in education, training, and knowledge exchange, stakeholders can ensure that everyone involved has the skills and understanding necessary to contribute meaningfully to mangrove conservation. The success of such collaborations hinges on the willingness to work together, respect local knowledge, and share resources.

7.6 Strategic Recommendations for the Future

Looking ahead, there are several strategic recommendations for the future of mangrove conservation in Africa. First and foremost, there is a need to strengthen policy frameworks that prioritize mangrove protection and restoration. Governments across Africa should commit to including mangrove ecosystems in national and regional climate adaptation strategies and development plans. This can be achieved through the integration of mangrove conservation into broader environmental policies, such as biodiversity protection, climate change mitigation, and sustainable development.

Fostering greater community engagement and participation in mangrove management is another key recommendation. Community-led restoration projects, such as those outlined in successful case studies, should be expanded and supported. Additionally, local knowledge and practices should be integrated into scientific research and policy-making, ensuring that conservation strategies are both culturally relevant and scientifically sound.

There is also a pressing need for increased investment in research, monitoring, and data collection to enhance our understanding of mangrove ecosystems and their contributions to climate resilience. This research can inform better management practices and help stakeholders track progress toward conservation goals. It will also help quantify the economic benefits of mangroves, providing evidence for policymakers to justify increased investment in these critical ecosystems.

7.7 The Path Toward Building Climate-Resilient Communities Along Africa's Coasts

In the face of rising climate risks, the path to building climate-resilient communities in Africa's coastal regions lies in the protection and restoration of mangrove ecosystems. By recognizing the interconnectedness of people and nature, Africa's coastal regions can embrace a sustainable future that prioritizes both environmental and socio-economic resilience. As this book has illustrated, mangroves are indispensable in the fight against climate change, serving as nature's solution to many of the challenges faced by coastal communities.

The resilience of these communities depends on safeguarding their natural resources, and mangrove conservation is central to that goal. By investing in the protection and restoration of these ecosystems, we not only ensure the survival of the mangroves themselves but also the well-being of the millions of people who depend on them. Long-term resilience will require a collective effort, with governments, NGOs, local communities, and the private sector all playing a role.

7.8 Call to Action

The time for action is now. Mangrove conservation must be prioritized as a critical strategy for building climate resilience along Africa's coasts. Governments, NGOs, local communities, and the private sector must come together to protect these vital ecosystems, ensuring that they continue to provide the many services that coastal communities rely on. By working collaboratively, sharing resources, and investing in both environmental and socio-economic sustainability, Africa's coastal regions can chart a path toward a more resilient and sustainable future for all.

Through these collective efforts, we can safeguard the future of Africa's coastal communities, ensuring that they remain resilient in the face of climate change, while also preserving the invaluable mangrove ecosystems that are central to their survival and prosperity. The time to act is now, and the future of Africa's coasts depends on our commitment to protecting these vital ecosystems.

Index

A

Abidjan Convention 17, 135, 136
Adaptation 19, 21, 44, 68, 109, 115, 140, 155
Adaptation strategies 20, 43, 47, 69, 92, 93, 104, 116, 138
Africa 2–8, 10, 12, 14, 15, 17–20, 22, 32, 34, 35, 39, 41, 43–46, 124, 129, 134, 152–158
Agriculture 2, 3, 5, 9, 16, 17, 19, 20, 39, 40, 56, 57, 69, 80, 82, 90, 91, 124, 139
Angola 102, 103, 107, 110, 115
Artisanal fisheries 59
Artisanal fishing 6, 8, 80
Atlantic Ocean 102, 124

B

Biodiversity 3, 5, 6, 8, 9, 14, 18, 20, 32–34, 40–42, 46, 47, 56–61, 63, 65, 66, 68–70, 80–82, 84, 86, 90, 102, 103, 106, 108–110, 112, 113, 116, 124–126, 128, 129, 132–134, 138, 140, 152
Biodiversity conservation 4, 15, 18, 43, 45, 56, 61, 64, 88, 136, 141
Biodiversity preservation 129
Biodiversity protection 156
Blue carbon 114

C

Cameroon coastal zone 56
Capacity building 19, 45, 69, 70, 87, 89, 90, 92, 93, 155, 156

Index

Carbon sequestration 4, 15, 20, 45, 47, 56, 57, 59, 63, 64, 85, 88, 90, 109, 112, 116, 124, 127, 129, 138–140, 152
Carbon sequestration by mangroves 81, 114, 153
Carbon sinks 18, 59, 68, 86, 103, 115, 125, 153
Casamance region 19, 131, 135
Case studies 14, 15, 20, 21, 43, 47, 154, 156
Central Africa 14, 20, 21, 56–63, 65–70
Central Africa coastal zones 56
Climate adaptation strategies 18, 19, 109, 156
Climate change 2, 4–8, 10, 11, 14, 15, 18, 20, 21, 32–34, 41, 43–45, 47, 57, 59, 63, 68, 70, 81, 85, 88, 90–93, 103, 104, 109, 111–117, 125, 127, 129, 136, 138–140, 152, 153, 156–158
Climate change adaptation 43, 47, 69, 92, 93, 104, 138
Climate change and coastal erosion 2, 8, 47
Climate change impacts 4, 8, 10, 14, 15, 47, 68, 70, 85, 88, 90, 103, 104, 115, 117, 125, 127, 152
Climate change mitigation 21, 33, 44, 68, 114–116, 156
Climate change resilience 42, 103, 125
Climate resilience 18, 19, 22, 47, 115, 140, 152, 153, 156, 157
Climate-resilient communities 157

Coastal areas 10, 16, 33, 37, 40, 56, 60, 68, 70, 80, 81, 86, 102, 105, 109, 110, 124–127, 133, 155
Coastal biodiversity 108
Coastal challenges 2, 20, 21, 80
Coastal communities 2–5, 7, 8, 11, 13, 15, 18, 33, 34, 39, 44, 57, 59, 64, 66, 68, 70, 81, 82, 85, 88, 103, 104, 109, 111, 112, 116, 132, 140, 152, 154, 155, 157
Coastal communities and livelihoods 6, 67, 81
Coastal degradation 3, 33, 85, 103
Coastal development 6, 8, 40, 41, 45, 47, 66, 115, 129, 140
Coastal ecosystems 3, 6, 7, 21, 33, 34, 39, 43, 45, 48, 59, 65–67, 82, 85, 91, 102–104, 110, 112–115, 117, 126, 135, 138, 139
Coastal erosion 2, 4, 8, 17, 19, 32–34, 36, 39, 40, 42, 44, 47, 59, 83, 85, 87, 92, 103, 104, 113, 115, 116, 125, 128, 131, 133, 139, 140, 152, 153
Coastal migration 81
Coastal protection 8, 15, 19, 39–41, 56, 57, 59, 61, 63, 64, 88, 90, 105–107, 109, 110, 112, 124, 128, 129, 133, 136, 138, 139
Coastal protection with mangroves 111, 133, 137
Coastal region sustainability 62, 126, 157
Coastal resilience 4, 10, 20, 47, 86, 104, 152
Coastal wetlands 102, 128, 132

Coastal zone management 46, 47, 91, 137, 138
Collaboration 14–16, 20, 42, 43, 64, 70, 91, 93, 113, 129, 130, 135, 137, 138, 140, 154, 156
Community-based conservation 42, 92, 128
Community-based forestry 67
Community-based initiatives 81, 86
Community-based management 16, 47, 62, 89, 91, 109
Community-based natural resource management (CBNRM) 12, 137
Community-driven monitoring 89
Community-driven solutions 152
Community engagement 14, 63, 64, 70, 87, 135, 136, 156
Community involvement 14, 43, 47, 64, 87, 89, 116, 129, 154
Community leadership 45
Community-led conservation 48, 62
Community participation 41, 47, 64, 114, 130
Community resilience 4, 14, 20–22, 43, 45, 70, 92, 93, 116, 140, 153
Construction materials 62, 153
Coral reefs 6, 32, 59, 80, 102, 103
Cross-border conservation 91
Cross-sectoral partnerships 22, 155

D

Deforestation 6, 12, 15, 19, 21, 42, 57, 63, 66, 68, 70, 82, 88, 92, 114, 125, 129, 136

E

East Africa 11, 13, 21, 80, 82, 83, 85, 86, 88–90, 92, 93
Economic growth 17, 66, 125
Economic value 21, 69, 111, 154
Ecosystem-based adaptation (EbA) 45, 68, 69, 112, 140
Ecosystem services 18, 34, 65, 68, 70, 82, 92, 109, 124, 132, 136, 138, 139
Ecosystem services of mangroves 17
Ecotourism 7, 8, 40, 67, 69, 81, 87, 88, 109, 111, 113, 116, 127
Empowerment 12, 20
Environmental degradation 4, 8, 44, 57, 81, 125
Environmental impact assessments (EIAs) 17, 67
Environmental regulations 127
Environmental sustainability 154
Environmental threats 2, 154
Equatorial Guinea coastal region 56

F

Firewood 42, 139, 153, 155
Fisheries 4, 6–9, 16, 18, 20, 32, 33, 35, 40, 47, 56, 57, 59, 80, 82, 88, 92, 102, 107, 109, 111, 113, 116, 124, 128, 129, 131–133, 135, 136, 139, 140, 152–155
Fisheries management 4, 19
Fishery resources 59, 68
Fishing industry 40, 42, 56, 80, 124, 125, 153
Flooding 2, 3, 5, 7, 18, 19, 21, 34, 59, 65, 81, 85, 103, 104, 125, 127, 152, 153

Fuelwood 14, 19, 66, 68, 82, 103, 127, 136

G
Gabon coastal region 20, 56, 68
Global warming 70, 81, 114, 115
Governance 16, 135, 155

H
Habitat fragmentation 67

I
Indian Ocean 80, 102, 112, 113
Indigenous knowledge 12, 61, 62, 91
Industrialization 3, 20, 21, 33, 57, 66, 67, 126, 127
Integrated coastal management (ICM) 103, 114
Integration of traditional ecological knowledge 14, 62, 154

K
Kenya 3, 7, 8, 11, 13, 14, 21, 80–82, 86, 88–92
Keta lagoon 129, 133

L
Land reclamation 15, 63, 86
Legal frameworks 67, 69, 87, 115, 135, 138
Livelihoods 5–7, 9, 10, 12–15, 18–22, 32, 35, 39–45, 48, 57, 59, 61, 63, 64, 67–70, 80–82, 86–88, 92, 103, 108, 111–113, 116, 124–127, 131, 135–137, 139, 140, 153, 154
Local ecological knowledge (LEK) 11, 62
Local knowledge 11, 12, 15, 16, 19, 44, 47, 63, 69, 88, 89, 128, 136, 152, 156
Local livelihoods 8, 10, 11, 20, 39, 41, 43, 82, 110, 116, 128, 152
Local stewardship 13, 48, 62
Logging 8, 57, 63, 64, 66, 67, 69, 90, 136

M
Madagascar 3, 102–105, 109, 110, 114, 115
Mangrove biodiversity 65
Mangrove conservation 6, 12, 14–17, 20–22, 44–48, 62–65, 69, 70, 86, 87, 89, 91–93, 113, 115, 116, 135, 136, 138–140, 152, 154–157
Mangrove conservation initiatives 12, 86, 129, 135
Mangrove conservation policies 46, 47
Mangrove ecosystems 6–12, 14, 17–22, 34, 35, 38–47, 57, 59–63, 66–71, 81, 82, 84, 86, 88–93, 104, 107–116, 126–129, 136, 137, 139, 140, 152–158
Mangrove ecosystems in Africa 10, 11, 15, 20
Mangrove forests 6–8, 13–17, 19, 20, 33, 34, 36, 38, 40, 42, 43, 45, 46, 48, 56–61, 63–68, 70,

Index

81, 82, 84, 87, 90, 91, 102–104, 106, 109–113, 115, 116, 124–129, 132–136, 138, 139, 153–155
Mangrove rehabilitation 46, 63, 65
Mangrove rehabilitation projects 65
Mangrove resource exploitation 62
Mangrove restoration 6, 7, 10, 13, 14, 17–21, 33, 41–44, 64, 67, 68, 90, 91, 93, 104, 109, 112, 114–117, 135, 137, 138, 140, 141, 154–156
Mangrove Restoration Initiatives 9, 12, 19, 89, 91
Mangroves 3–8, 10, 11, 13–22, 33–35, 38–48, 56–64, 66–70, 80–83, 85, 86, 88–92, 102–104, 108–117, 125–140, 152–155, 157
Mangrove species 11, 36, 38, 39, 59, 60, 62, 83–85, 105, 109, 128, 131, 133
Mangrove species replanting 13, 67
Marine biodiversity 6, 102, 153
Marine protected areas (MPAs) 46, 139
Marine resources 6, 8, 32, 40, 80, 81, 108
Medicinal plants 62, 70, 103, 112, 153, 155
Mozambique 3, 6, 8, 12–14, 21, 82, 102–104, 109–115
Mozambique mangrove conservation 14

N
Namibia 21, 102

Nationally Determined Contributions (NDCs) 115, 140
Niger Delta 3, 125–127, 129, 134
Non-timber forest products 140, 155

O
Océanium 130
Oil exploration 3, 21, 66, 125, 126
Oil spills 3, 57, 66, 125–127
Overexploitation 32, 33, 62, 69, 92, 112, 136, 152
Overfishing 2, 3, 11, 21, 32, 40, 80, 81

P
Participatory mapping 89
Payments for Ecosystem Services (PES) 16, 92
Policy development 69, 113, 155
Policy frameworks 4, 15, 20, 69, 87, 90, 129, 135, 136, 140, 156
Pollution 6, 8, 15, 32, 45, 48, 70, 86, 115, 116, 125–127, 135, 152
Private sector involvement 139

R
Red sea coast 20, 32, 34–46, 48
Reforest'Action 129
Regional collaboration for mangrove restoration 20, 21
Regional cooperation 16, 17, 20, 21, 46, 47, 91, 113, 116, 135, 155
Research and monitoring 48

Resilience 7, 18, 21, 33, 39, 44, 48, 65, 68, 70, 71, 81, 85, 87, 104, 109, 112, 116, 139, 157
Resilience building 68
Restoration projects 10, 13, 19, 20, 22, 41, 42, 64, 65, 86, 91, 109, 112, 115–117, 129, 135, 139, 154, 156
Rising sea levels 2, 4, 5, 7, 8, 18, 34, 44, 59, 85, 103, 104, 111, 115, 116, 125, 140, 152

S

Scientific research 12, 15, 21, 44, 62, 70, 88, 114, 140, 156
Seagrass meadows 102
Sea-level rise 32, 68, 91, 140
Seawater Solutions (SwS) 129
Sediment stabilization 38, 60, 105, 107, 109, 110
Senegal mangrove restoration 135, 139
Senegal Mangrove Restoration Project 130
Seychelles 80, 90, 92
Sine-Saloum Delta 135
Socio-economic pressures 2, 3
Somalia 21, 80, 81, 86, 87, 90
Songhor lagoon 129
South Africa 21, 102–104, 107, 109, 110, 112, 113, 115
Southern Africa 21, 102–105, 109–117
Storm frequency 152
Storm protection 20
Storm surge mitigation 7, 68
Storm surges 3, 8, 10, 18, 34, 37, 44, 47, 57, 59, 68, 70, 81, 83, 85, 104, 111, 115, 116, 125, 140, 153
Sustainable development 4, 22, 33, 117, 126, 141, 156
Sustainable fishing practices 43, 64
Sustainable logging practices 67
Sustainable management 7, 15, 17, 19, 21, 44–46, 69, 70, 88, 90, 92, 129, 138, 140, 155
Sustainable mangrove harvesting 14
Sustainable practices 61, 67, 70, 88, 92, 113, 116, 137, 139, 152, 154
Sustainable resource management 62, 64, 67, 86, 112, 116, 135
Sustainable resource use 45, 62, 86

T

Tanzania 6, 8, 11, 21, 80, 82, 86, 88, 90–92, 112, 115
Tidal waves 18, 153
Timber exploitation 66
Tourism 2, 3, 5, 8, 32, 39–41, 56, 80, 82, 91, 110, 111, 124, 139
Traditional ecological knowledge (TEK) 44, 137, 139
Traditional ecological practices 44
Traditional knowledge 10, 12, 20, 21, 43, 44, 48, 61, 70, 88, 89, 137, 138, 154
Transboundary mangrove management 16
Tropical cyclones 104

Index

U

United Nations Sustainable Development Goals (SDGs) 92
Urbanization 3, 11, 19, 21, 63, 66, 88, 116, 125–127, 152

V

Vulnerability 2, 3, 19, 21, 57, 68, 85, 103, 104, 109, 115, 125, 127

W

Water filtration 35
West Africa 2, 10, 13, 21, 22, 58, 124–132, 135–138, 140, 141, 154
West Africa Marine Ecoregion (WAME) 135
Wetlands 40, 46, 91, 124, 125

Z

Zambezi Delta 113
Zanzibar Association for Climate Change Resilience (ZACCR) 87, 90

The manufacturer's authorised representative in the EU is Springer Nature Customer Service Centre GmbH, Europaplatz 3, 69115 Heidelberg, Germany. If you have any concerns regarding our products, please contact ProductSafety@springernature.com

Printed and bound by CPI Group (UK) Ltd, Croydon, CR0 4YY
26/03/2026
02078942-0005